USING THE STRUCTURED TECHNIQUES
A Case Study

YOURDON PRESS COMPUTING SERIES
Ed Yourdon, *Advisor*

BENTON AND WEEKES   Program It Right: A Structured Method in BASIC
BLOCK   The Politics of Projects
BODDIE   Crunch Mode: Building Effective Systems on a Tight Schedule
BRILL   Building Controls Into Structured Systems
BRILL   Techniques of EDP Project Management: A Book of Readings
CONSTANTINE AND YOURDON   Structured Design: Fundamentals of a Discipline of Computer Program and Systems Design
DE MARCO   Concise Notes on Software Engineering
DE MARCO   Controlling Software Projects: Management, Measurement, and Estimates
DE MARCO   Structured Analysis and System Specification
DICKINSON   Developing Structured Systems: A Methodology Using Structured Techniques
FLAVIN   Fundamental Concepts in Information Modeling
HANSEN   Up and Running: A Case Study of Sucessful Systems Development
KELLER   Expert Systems Technology: Development and Application
KELLER   The Practice of Structured Analysis: Exploding Myths
KING   Current Practices in Software Development: A Guide to Successful Systems
KRIEGER, POPPER, RIPPS, AND RADCLIFFE   Structured Micro-Processor Programming
MAC DONALD   Intuition to Implementation: Communicating About Systems Toward a Language of Structure in Data Processing System Development
MC MENAMIN AND PALMER   Essential Systems Analysis
ORR   Structured Systems Development
PAGE-JONES   The Practical Guide to Structured Systems Design
PETERS   Software Design: Methods and Techniques
ROESKE   The Data Factory: Data Center Operations and Systems Development
SEMPREVIO   Teams in Information Systems Development
THOMSETT   People and Project Management
WARD   Systems Development Without Pain: A User's Guide to Modeling Organizational Patterns
WARD AND MELLOR   Structured Development for Real-Time Systems, Volumes I, II, and III
WEAVER   Using the Structured Techniques: A Case Study
WEINBERG   Structured Analysis
WELLS   A Structured Approach to Building Programs: BASIC
WELLS   A Structured Approach to Building Programs: COBOL
WELLS   A Structured Approach to Building Programs: Pascal
YOURDON   Classics in Software Engineering
YOURDON   Coming of Age in the Land of Computers
YOURDON   Design of On-Line Computer Systems
YOURDON, LISTER, GANE, AND SARSON   Learning to Program in Structured Cobol, Parts 1 and 2
YOURDON   Managing Structured Techniques, 3/E
YOURDON   Managing the System Life Cycle
YOURDON   Structured Walkthroughs, 2/E
YOURDON   Techniques of Program Structure and Design
YOURDON   Writing of the Revolution: Selected Readings on Software Engineering
ZAHN   C Notes: A Guide to the C Programming

USING THE STRUCTURED TECHNIQUES
A Case Study

Audrey M. Weaver

YOURDON PRESS
A Prentice-Hall Company
Englewood Cliffs, NJ 07632

© 1987 by Prentice-Hall, Inc.
A Division of Simon & Schuster
Englewood Cliffs, New Jersey 07632

All rights reserved. No part of this book may be
reproduced, in any form or by any means,
without permission in writing from the publisher.

Printed in the United States of America

10 9 8 7 6 5 4 3 2 1

ISBN 0-13-940263-2   025

Prentice-Hall International (UK) Limited, London
Prentice-Hall of Australia Pty. Limited, Sydney
Editora Prentice-Hall do Brasil, Ltda., Rio de Janeiro
Prentice-Hall Canada Inc., Toronto
Prentice-Hall Hispanoamericana, S.A., Mexico
Prentice-Hall of India Private Limited, New Delhi
Prentice-Hall of Japan, Inc., Tokyo
Prentice-Hall of Southeast Asia Pte. Ltd., Singapore

In memory of my father

Acknowledgments

I would like to express my thanks to my good friends and coworkers, Jill Knight and Bob Moss, for being sounding boards for the ideas which led to this book; Julia Tinsley for her valuable suggestions and thorough review of the manuscript; my husband, Jim, and sons, Ralph and Alan, for their patience and support; the students in my classes who often "had a better idea"; Indiana Bell for providing a work environment where I have been given the opportunity to put my ideas into practice; and Carol Crowell for copyediting of the manuscript.

– Audrey Weaver

CONTENTS

INTRODUCTION		**xi**
HISTORY OF STRUCTURED TECHNIQUES		xi
Structured Coding		xi
Structured Design		xii
Structured Analysis		xiii
Information Modeling		xiv
PROBLEMS WITH LEARNING STRUCTURED TECHNIQUES		xiv
SUGGESTED USE OF THIS TEXTBOOK		xv
AUDIENCE FOR THIS TEXTBOOK		xvi
REFERENCES		xvii
CHAPTER 1	**THE PROJECT ASSIGNMENT**	**1**
	QUESTIONS	4
	REFERENCES	4
CHAPTER 2	**THE PROJECT SCOPE**	**5**
	IDENTIFYING THE PROBLEM, OBJECTIVES, AND EVENTS	5

DETERMINING SYSTEM INPUTS, OUTPUTS AND USERS	6
COLLECTING SAMPLE FORMS AND DOCUMENTS	9
QUESTIONS	12
REFERENCES	12

CHAPTER 3 DATA DICTIONARY 13

QUESTIONS	18

CHAPTER 4 THE INFORMATION MODEL 19

OBJECTS AND RELATIONSHIPS	21
IDENTIFYING OBJECTS	24
ATTRIBUTING DATA ITEMS TO OBJECTS	27
QUESTIONS	34
REFERENCES	34

CHAPTER 5 THE REFINED INFORMATION MODEL 35

REVIEWING THE FIRST-CUT INFORMATION MODEL	36
Eliminating Objects	37
Eliminating Multiple Relationships	38
Adding an Associative Object	38
Attributing Derived Data	39
THE REFINED INFORMATION MODEL	41
QUESTIONS	45
REFERENCES	45

CHAPTER 6 THE PROCESS MODEL 46

DIAGRAMMING EVENTS	46
Event 1	47
Event 2	48
Event 3	48
Event 4	49
DRAWING A CONSOLIDATED EVENT DIAGRAM	50
QUESTIONS	54
EXERCISE	54
REFERENCES	54

CHAPTER 7 THE REFINED PROCESS MODEL   **55**

    WALKING THROUGH THE EVENT DIAGRAM   55
    DESCRIBING THE MAJOR PROCESSES   59
    DEFINING INFORMATION REQUIREMENTS   62
    CHANGING PLANS   64
    USING A FOURTH GENERATION LANGUAGE   65
    QUESTIONS   66
    EXERCISES   66
    REFERENCES   66

CHAPTER 8 THE DETAILED PROCESS MODEL   **67**

    EXPLODING THE DATA FLOW DIAGRAM   67
    Design Curriculum   68
    Schedule Training   71
    Train Students   77
    Maintain Personnel Information   83
    HOLDING A WALKTHROUGH   88
    MAKING REVISIONS   88
    PROCESS ALLOCATION   90
    QUESTIONS   90
    EXERCISES   91
    REFERENCES   91

CHAPTER 9 THE DESIGN MODEL   **92**

    DRAWING THE EXPANDED DATA FLOW
      DIAGRAM   94
    CONSTRUCTING THE STRUCTURE CHART   95
    EVALUATING THE STRUCTURE CHART   102
    CORRECTING AN OMISSION   103
    QUESTIONS   104
    EXERCISES   105
    REFERENCES   105

CHAPTER 10 THE PACKAGED MODEL   **106**

    GROUPING PROCESSES INTO MODULES   107
    GROUPING MODULES INTO PROGRAMS   107
    USING PROCEDURAL ANALYSIS   108
    TUNING FOR EFFICIENCY   112
    DESIGNING THE CODE   112
    QUESTIONS   113

	EXERCISES	113
	REFERENCES	113
CHAPTER 11	**THE MAINTENANCE REQUEST**	**114**
	ANALYZING THE CHANGE	115
	REVISING THE INFORMATION MODEL	116
	REVISING THE PROCESS MODEL	117
	QUESTIONS	118
	EXERCISE	119
	REFERENCES	119
APPENDIX A	**MANUAL SYSTEM SAMPLE FORMS AND DOCUMENTS**	**120**
APPENDIX B	**REQUIREMENTS DEFINITION**	**126**
APPENDIX C	**DETAILED PROCESS MODEL**	**148**
APPENDIX D	**DESIGN MODEL— MAINTAIN PERSONNEL INFORMATION**	**157**
APPENDIX E	**DATA DICTIONARY**	**186**
APPENDIX F	**ANSWERS TO QUESTIONS AND EXERCISES**	**192**
	GLOSSARY OF TERMS	**225**
	GLOSSARY OF SYMBOLS	**231**
	INDEX	**237**

Introduction

I conclude that there are two ways of constructing software design: One way is to make it so simple that there are obviously *no deficiences and the other way is to make it so complicated that there are no* obvious *deficiencies.*
C.A.R. Hoare, "The Emperor's Old Clothes," *Communications of the ACM*, Vol. 24, No. 2, February 1981.

Although structured techniques are the standard for software development in many companies today, few analysts have developed real competence in applying them. When a project team tries to develop a system using structured techniques, the analysts flounder and waste valuable project time. I believe the reason that they have trouble is rooted in the history of the development of structured techniques.

When I first started programming in the mid-1960's, we had not yet encountered the maintenance load that is bogging down data processing shops today. The only thing that mattered in those earlier days was to write programs that worked correctly and did not "bomb" too often in the middle of the night. Just getting the input cards through the card reader was a major accomplishment. It wasn't until several generations of programmers made changes to these programs that we realized we also had to write systems that minimized maintenance. Somehow we needed to get some order, some structure, into the software development process.

History of structured techniques

Structured coding

The first of the structured techniques was structured programming, or more correctly, structured coding. The theory was that if all program code were written using only three structures — sequence, selection, and iteration — there would be fewer errors, and a programmer could change any other programmer's code with greater speed and accuracy.

Books about structured programming were written, courses developed, and all enlightened data processing shops instituted standards stating, "Thou shalt write structured code." To a large extent, programmers thought that this dictum made sense and indeed did write structured code and there were no more maintenance problems. Well, not quite.

What actually happened was that even though most programmers gradually started writing structured code and the programs became easier to maintain, severe maintenance problems still occurred. The software thinkers of those days soon realized that even though structured code was good, unnecessary maintenance still was occurring because every change in specifications triggered code changes throughout the system.

Structured design

The solution was to design systems so that each section of code contained one and only one function, each section was as independent as possible from other sections, and sections were small enough so that the programmer could comprehend the entire section at once. Redundant code was eliminated and the sections were organized in levels much like an organization chart. Structured design was born.

Rules were formulated, techniques developed, books written, courses taught, competing schools of thought debated, and most enlightened data processing shops incorporated some type of structured design into their standards. Designers all started using structured design and the maintenance problems went away. Well, no.

First, since there were several competing schools of thought regarding the best form of structured design, some data processing shops chose one method, some chose another, and some decided it was too confusing and ignored the whole issue. Even in those shops that chose a method, instituted standards, and trained their people, it took a long time for structured design techniques to begin to make an impact. One reason was that while nearly everyone coded, only those people involved in new development or enhancements had a chance to design. Often their first chance to try structured design was on an important visible project and they were not allowed time for the learning curve.

Where structured design was given a chance, it did indeed make systems easier and quicker to maintain, but the maintenance effort still consumed most of a data processing shop's resources.

After structured design had been used for a while, the software thinkers realized that one of the main reasons so much maintenance was required was that the developmental specifications often were incomplete. The changes we called maintenance largely were functions that should have been in the system from the beginning. The business functions needed to be more rigorously analyzed before the mechanized system was designed.

Structured analysis

Several techniques were developed to aid in the analysis process. Data flow diagrams depicted the flow of data through the system, process descriptions (mini-specifications) used unambiguous prose to describe what happened to the data, and a data dictionary precisely defined the data. The documentation was developed in levels, much like an outline, making both development and use easier. Structured analysis was born.

Rules were formulated, techniques developed, books written, courses taught, and enlightened data processing shops included structured analysis in their development standards. All analysts used structured analysis, and maintenance took only its fair share of resources. Not quite.

Some analysts thought that they could produce a structured design just fine without bothering with structured analysis. Many analysts thought structured analysis made sense, but encountered the same problems as with structured design. Again, there were some competing methods (the differences were not as great as with the design methods), but the main problem was still that they did not get a chance to try the technique until they were on an important project, which is a very bad time to try to learn a new technique.

Where structured analysis was tried, another problem quickly developed. Most analysts had a hard time trying to figure out how to go from structured analysis to structured design because the structured design books were written prior to the structured analysis books. Once analysts began to get some experience with structured analysis and structured design, they realized that there is a smooth and natural transition from one to the other. But since the techniques largely were developed and taught independently, many people never learned to use the techniques smoothly.

Structured analysis helped a lot in getting a complete understanding of the problem before designing the solution. Analysts were defining the processes and the data flowing through the system, but still were having trouble getting the data elements assigned to data stores (data at rest), a step which would lead to well-designed files and databases. (Conversely, the design techniques which concentrated on analyzing the data had problems with the design of the processes and data flows.)

Information modeling

Software thinkers began to realize that the solution was to define the data stores first. This step could be done by identifying the things about which an organization stores data and grouping the data elements according to what they describe. Information modeling was born.

Books were written, techniques developed, courses taught, and all enlightened data processing shops incorporated information modeling into their standards. All system developers developed information models and the maintenance problem was solved. Well, no.

As with structured design, several competing methodologies are being promoted, along with different terms for each method. Now people not only do not understand how to move from structured analysis to structured design, but also they do not know how to move from information modeling to structured analysis. In addition, they still have to learn to use the techniques on their first big project.

Problems with learning structured techniques

As I reflected on this history of the development of structured techniques, I realized that because the techniques were learned individually, analysts did not understand how the whole process should flow until they had a chance to develop a system from start to finish. No wonder projects were taking too long to develop. Everyone was learning on the job. Only those of us who had the good fortune to get a second and third try were becoming competent in using the techniques.

Solving this problem is not easy. Because the major books on structured techniques were published as the techniques were evolving, they inadvertently added to the problem. Most data processing people are maintenance programmer/analysts who may get a chance to do a little structured analysis here and a little structured design there. While use of any one of the structured techniques by itself produces a better system which requires fewer maintenance resources, use of all of them as a cohesive approach produces the really significant benefits. *What was needed was a course to give the student practice in doing the complete process from start to finish.*

When I started teaching structured techniques a few years ago, I also realized that most examples in the textbooks were just pieces of larger problems. Often the examples in a book were even from several different applications. My students complained that this made the techniques very difficult to learn. *What was needed was a course in which all the examples were from the same application.*

Another complaint my students had was that when a textbook did have a complete case study it took so much effort to understand the business system being studied that they couldn't concentrate on learning the structured techniques. *What was needed was a case study about some very familiar business function.*

Many courses and texts on structured techniques do not fit the techniques into the framework of a system development life cycle. The students then have problems trying to decide in which phase each technique should be used. *What was needed was a textbook that taught structured techniques within a typical system development life cycle.*

Suggested use of this textbook

This book was developed to fill the need for a textbook on structured techniques which solves all four problems. It guides the student through information modeling, structured analysis, structured design, and the design of structured code. It uses the same case study throughout the entire text. The application studied — training record-keeping — is one that is familiar to all students, and the book follows a typical development life cycle.

Not only is the training record-keeping application one that is familiar to students, but also it is just the right size for a case study. It contains a complete set of functions, yet is small enough that the entire application can be analyzed. The system breaks down into two subsystems, one of which is designed in the case study.

This text is not intended for use as a stand alone text on structured techniques. The assumption is made that the individual techniques already have been studied, and this case study shows the student how the various techniques fit together within a development life cycle.

Chapters 1 through 3 discuss the nature and scope of the project and the initial data gathering activities. In Chapters 4 and 5, an information model is used to analyze the system data. In Chapters 6 through 8, structured analysis is used to analyze the system processes. In Chapters 9 and 10, structured design is used to design the internal structure of the system. Chapter 11 provides an example of making a maintenance change to a system which was developed using structured techniques. Questions have been provided at the end of each chapter to test the readers' understanding of the text. Also, starting in Chapter 6, exercises involving a simple case study are provided to give readers an opportunity to practice the structured analysis and design techniques they have just learned. A Glossary of Symbols explaining the diagramming techniques is included, as well as a Glossary of Terms.

Because my students constantly ask me where they can find examples of development documentation, the appendices include the Requirements Definition document, the Detailed Analysis document, and a partial Design document. The appendices also include sample forms and documents from the manual system, the data dictionary, and answers to the questions and exercises which are at the end of each chapter.

The case study discusses mechanizing an existing manual system rather than the more difficult problem of rewriting or enhancing an old mechanized system. *Essential Systems Analysis* by John Palmer and Stephen McMenamin thoroughly examines rewriting existing mechanized systems.

Because the text focuses on the use of structured techniques in developing the internal design of a system, many design issues are not addressed in the text — issues such as determining whether a system should be batch or on-line, distributed versus centralized processing, hardware requirements, and screen design. An excellent book on design is *The Design of Complex Information Systems* by Grayce M. Booth. Neither logical nor physical database design is covered in this text. One of the better books on database design is *Introduction to Database Systems*, 3rd edition, by Chris Date.

Audience for this textbook

This book primarily is intended for the inexperienced programmer/analyst or second or third year college student. After completing this case study, the student should be able to develop a small system or a piece of a large system and understand how it fits into the total development process.

This book is also for maintenance programmers s⟨
are put on that long-awaited development project they
had practice using structured techniques as a cohe
methodology. In addition, it will help them maintain
were developed with structured techniques.

Finally, anyone who works with programmers
as data processing managers, users, and software
find the case study useful in increasing their unde
tem development process.

References

Booth, Grayce M. *The Design of Complex*
 York: McGraw-Hill, 1983.

Date, Chris. *Introduction to Database Sy⟨*
 Addison-Wesley, 1981.

Martin, James and Carma McClure. *Stru⟨*
 Englewood Cliffs: Prentice-Hall, Inc.

Palmer, J. and S. McMenamin. *Essent*
 YOURDON Press, 1984.

USING THE STRUCTURED TECHNIQUES
A Case Study

1
The Project Assignment

> ... *I believe in starting with an attempt to understand and accept the state of the world today* — *the way things are, rather than the way I would like them to be* . . . . *Next, I must have some idea of where to land at the end of my great leap. I must discover what we would like things to be. And before I can do that, I must discover who "we" are.*
> Gerald M. Weinberg, *Rethinking Systems Analysis and Design.*
> Boston: Little, Brown and Company, 1982, p. 144.

Alan obtained a degree in computer information systems and six months ago began working at International Telewidgets Corporation. He has been learning about the company, has been trained to use the development tools provided for the programmers and analysts, and has had a few maintenance assignments. Now his manager has a development request that is the right size for Alan's first project.

The data processing training coordinator, Terri, has asked if they could please "do something" about mechanizing the data processing training records. Alan's manager gives Terri's telephone number to Alan, shows him the system development standards manual, and says he'll check with him in a week to see how he is getting along.

Alan goes back to his desk and starts reading the standards manual, trying to figure out how to start. International Telewidgets uses a seven-phase development methodology: Feasibility Study, Requirements Definition, Detailed Analysis, Design, Implementation, System Test, and Conversion.

Alan decides that he first must define the scope of his project. According to the standards manual, defining the scope includes documenting the following:

2 THE STRUCTURED SYSTEMS LIFE CYCLE: A CASE STUDY

- The problem definition
- Project objectives
- User needs
- Events to which the system must respond
- System users
- System inputs and outputs
- Any constraints

Because he knows nothing about the existing system except that there is a problem, Alan jots down some questions that he wants to ask Terri and calls her for an appointment. Her secretary tells him Terri has just left on vacation and won't be back for two weeks.

Alan wonders whether to tell his manager and see if he can get some small assignment to do while he is waiting, but instead he decides to spend the time studying. Because he wants to try information modeling on this project, he reads *Fundamental Concepts of Information Modeling* by Matt Flavin. His manager comes by and makes some remark about "not spending too much time studying," starts to leave, then stops, and says, "Since you seem to like to read, why don't you find out what this prototyping idea is all about. Maybe you could try that and skip some phases."

Alan is fortunate that his department has a library, so he reads all he can find about prototyping. He reads that a prototype is an actual working model of something. He can understand how a company would want to build a prototype of an airplane, or perhaps of a software system if there is no way to determine ahead of time how it would work; but he cannot see how this concept applies to mechanizing data processing training records.

As he reads further, however, he realizes that prototyping is a technique for showing the customer what the mechanized system will look like *before* the internal design and coding are done. It is used to help define the requirements so that the customer is more likely to be happy with the finished product, resulting in pleased customers and less maintenance. Some software developers have been using this technique on a limited basis for a long time. But now, much better tools are available for simulating screen interactions and report layouts. Alan decides that he will watch for a way to use this technique as he defines the requirements.

Two weeks have passed and Monday morning at 8:00 A.M. Alan calls Terri's office again. This time her secretary tells him that the earliest Terri can talk to him is Friday morning at 8:30. He makes the appointment and spends the next four days reviewing some of the structured analysis techniques in *Structured Analysis and System Specification* by Tom DeMarco.

He starts drawing a context diagram (Level 0 data flow diagram), but all that he knows right now is shown in Figure 1.1.

Figure 1.1.

Without the user he can't even start. Fortunately, it is Thursday, and he will be meeting with Terri tomorrow.

Questions

1. What would you expect to find in a system development standards manual?
2. What is the difference between software development and system development?
3. What system development phase is Alan skipping? Should he be? Why?
4. What is the purpose of the Requirements Definition phase?
5. Why is it important to formally define the scope of a project?
6. What is meant by software development tools?
7. When is true prototyping justified in business software development?
8. What do you think about the manager's comment regarding using prototyping and skipping some phases?
9. What are some tools you could use for prototyping?
10. What are some of the questions that Alan should ask the training coordinator in his first interview?
11. What is meant by the term *user*?
12. According to *Fundamental Concepts of Information Modeling*, upon whose work is information modeling based? Who is Ted Codd? How do his data analysis techniques differ from information modeling?

References

DeMarco, T. *Structured Analysis and System Specification.* New York: YOURDON Press, 1978.

King, David. *Current Practices in Software Development.* New York: YOURDON Press, 1984.

Flavin, M. *Fundamental Concepts of Information Modeling.* New York: YOURDON Press, 1981.

2

The Project Scope

> *Most problem projects can trace a great deal of their trouble to the fact that there was never a clear understanding between DP and the user about the project requirements and specifications.*
> Claude W. Burrill and Leon W. Ellsworth, *Modern Project Management.* Tenafly: Burrill-Ellsworth Associates, Inc., 1980, p. 74.

Friday morning finally comes, and Alan is seated in Terri's office.

"It's nice to meet you, Alan," Terri says. "I'm glad you've been assigned to do something about mechanizing our training records. I wish you had a little more experience, but this really isn't that big a system.

"We've been so busy developing the curriculum and courses and teaching that we've just been keeping our records on the word processing system.

"I'm sure you've already been here long enough to notice that our year-to-date training report doesn't always match the monthly reports; we sometimes lose track of enrollments and sometimes end up with two classes scheduled into the same room."

Identifying the problem, objectives, and events

"I'd like to ask you a few questions so that I can get a better understanding of the problem," Alan says. "If I understood you correctly, the main problem is that the training records are being kept manually on the word processor, and the information in the various reports are not consistent with each other. Is that the only problem?"

"Well, that and the fact that entering the same information more than once is very time-consuming. And then there's all the paper work involved with the registrations."

"So by mechanizing the records, you save money and serve your students and their managers better," Alan says.

"Exactly," Terri agrees. "I estimate we would eliminate two clerical positions if we could mechanize. Plus the time the students would save by not having to call us to straighten out their records."

"What events occur that your system must respond to?" Alan asks.

"What do you mean?"

"Well, an example I can think of would be when a student decides to take a course. Your area needs to respond by providing the training."

"Oh," Terri says. "Then another example would be that when the skill requirements for a job change or a new job is created, we have to change the curriculum. Is that what you mean?"

"Yes. Can you think of anything else?"

"Something that causes us a lot of work is that people change work groups. Since we organize the reports by group, this means we have to change our records a lot."

Determining system inputs, outputs, and users

While she is talking, Alan is sketching, making sure that Terri can see what he is doing. See Figure 2.1.

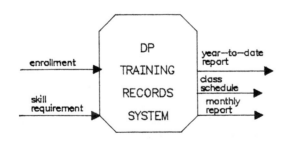

Figure 2.1.

"You mentioned that you send out monthly reports and year-to-date reports, and I've seen the class schedule you send out each month," Alan says. "Who prepares the enrollment forms?"

"The students or their managers. Haven't you taken any classes yet?" Terri asks.

"Yes, but my manager must have enrolled me. Who notifies you of changes in job skill requirements?"

"The managers."

"Who develops the curriculum?"

"I do."

"Who receives the monthly reports?"

"The students and their managers. Haven't you received any?" she asks him.

"No."

"Must be because you're new and we don't have you on our personnel list yet."

"How do you find out about personnel changes?" Alan asks.

"From the departmental telephone list."

"Oh, now I know why the training reports are not up to date. Who gets the year-to-date reports?"

"Students and managers."

"Are the monthly and year-to-date reports the same except for the time period they cover?"

"Yes."

"Who gets the course schedule?"

"Students, managers, and the registrar."

Alan's diagram now looks like Figure 2.2.

8 THE STRUCTURED SYSTEMS LIFE CYCLE: A CASE STUDY

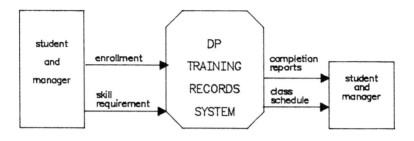

Figure 2.2.

"Any other reports that you produce?" Alan asks.

"Yes," Terri answers. "I have to give my boss a statistical report on training whenever he asks for one, but at least annually. Let's see. Oh, yes, we produce the Course Catalog, preliminary and final Class Rosters, and a Training Needs Forecast turnaround document."

"Do you report completions other than year-to-date and monthly?"

"Yes, on the Needs Forecast form. Also, we use the history to see if the student has completed course prerequisites."

"Any other features you'd like to see in a mechanized system?" Alan asks.

"I think we need an interactive method for updating the records and requesting the reports. We're giving the other departments all these new tools. I think we need some for ourselves now."

"I hate to rush, but I've got another appointment in five minutes. Do you have enough information to start?"

"Yes," Alan says. "I think so. Did I accurately sketch what we've been discussing?" He shows Figure 2.3 to Terri.

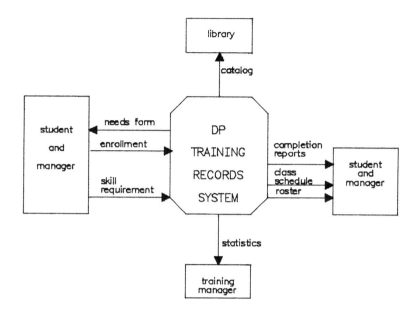

Figure 2.3.

Collecting sample forms and documents

"Yes, I believe so," Terri says, after studying the diagram. "By the way, I'd sure like the students to be able to get on the terminal and enter their own enrollments instead of sending in all those forms."

"Okay, I'll make a note of that. Could I have a copy of all the forms or reports that you currently use?"

"Certainly, but why do you want to see the reports we have now? Don't you want to know how I want my new reports to look?"

"Of course," Alan answers. "But we'll do that later. Right now, I just want to be sure that we store all the data that your application uses. After I make note of the data you currently use, we'll go over them together and see if there is anything that is no longer needed or anything we should add. The specific design of the new reports can be addressed fairly late in the design process."

"That makes sense. I'll have the clerk get them to you. Well, it's been nice meeting you. When will you have the new system done?"

"I won't know until I look at the forms and analyze your requirements. Then I'll get back with you. By the way, does the information you currently store and the reports you now produce enable you to keep track of everything you would like to?"

"Pretty much. It would be nice if a schedule could be produced for each room and each instructor also."

"I don't see why that couldn't be done. Well, thanks again. This looks like an interesting project."

As he is reviewing the notes from his interview with Terri, Alan starts thinking about the personnel changes. He wonders who makes up the telephone list and where the information comes from. After talking to several people, he discovers that the information comes from the department head's secretary, who gets the information from the personnel change requests after they are approved by the personnel department. Alan adds personnel information as an input coming from personnel and wonders how he is going to mechanize this input.

In a few days, Alan receives the documents from Terri's clerk (see Appendix A) and starts reviewing them. He is looking particularly for system inputs and outputs that Terri may not have mentioned. These in turn may indicate other users that he is unaware of.

He sees two reports that he isn't sure about. One is the Completion form and the other is the Forecasted Needs report. He calls Terri, who tells him that the Completion form is used just within the training area and that the Forecasted Needs report is produced from the needs indicated on the Needs Form turnaround document that is sent to the students. He adds them to his context diagram, which now looks like Figure 2.4.

Finally, Alan interviews selected students and managers to see if there is anything else that they need from the system.

The only additional need that he uncovers is that the students and their managers are upset that no-shows due to illness are reported in the same way as no-shows due to negligence. They would like to see the reason for the no-show listed on the reports. He also discovers that the no-show policy is either not well defined or not well understood.

THE PROJECT SCOPE 11

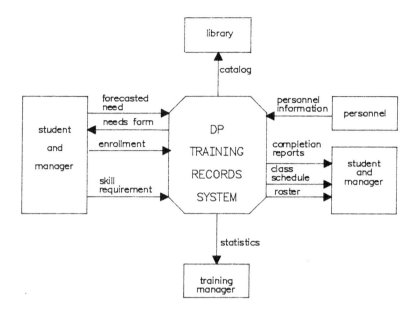

Figure 2.4.

What has Alan accomplished so far? He has learned who most of the users are, what events drive the system, what the major inputs and outputs are, the project objectives, and most of the user needs.

Questions

1. Why do you think Alan did not diagram the registrar as a user of the course schedule?
2. Why do you think Alan did not diagram the training coordinator as the source of the curriculum?
3. If the system did not exist already as a manual function, what would have to be done differently?
4. If the system already were partly mechanized, what additional information would Alan need to gather?
5. Write the problem definition.
6. What are the project objectives? (Objectives must be measurable.)
7. List the user needs defined so far.
8. List the events to which the system must respond.
9. Are there any constraints that we are aware of so far?

References

Burrill, C. W. and L. W. Ellsworth. *Modern Project Management.* Tenafly: Burrill-Ellsworth Associates, Inc., 1980.

King, David. *Current Practices in Software Development.* New York: YOURDON Press, 1984.

3

Data Dictionary

There's no consistent substitute for a thorough understanding of your problem, though sometimes people get lucky.
 Gerald M. Weinberg, *Rethinking Systems Analysis and Design.*
 Boston: Little, Brown and Company, 1982, p. 22.

Alan is sitting at his desk trying to decide what to do next. After a while, he starts looking in the system development standards manual. Thumbing through the chapter on requirements definition, he finds that the next step is to start the data dictionary. "That's odd," he thinks. "I thought the data dictionary was some software. I wonder what they mean."

He looks at the exhibit in the manual and sees that all that is required now is to identify each item of business data that the system needs and describe it in a sentence or two. International Telewidgets does have software that will store this information for him, but he could use something as simple as 5 x 7 cards if he had to.

The manual also suggests that a good way to start identifying this business data is by reviewing the forms and documents used by the existing system, being careful not to include extraneous data. Derived data is extraneous at this stage of analysis. In addition, he will need to include any new data that the system requires, but he will have to get that information from Terri.

Alan pulls out the stack of documents that Terri's clerk sent him and starts looking for business data items. On the January Course Roster, shown in Figure 3.1, he finds three data items: *course number* (C213), *course name* (Advanced Cobol), and *student name* (Brown B).

14 THE STRUCTURED SYSTEM LIFE CYCLE: A CASE STUDY

January in the report heading seems to indicate when the courses are scheduled, but it may be just when the report was printed.

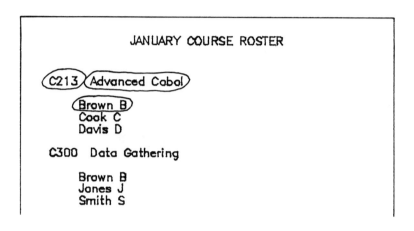

Figure 3.1.

From the Enrollment Request form he gets three more data items: *course start date, student number,* and *enrollment date,* shown in Figure 3.2.

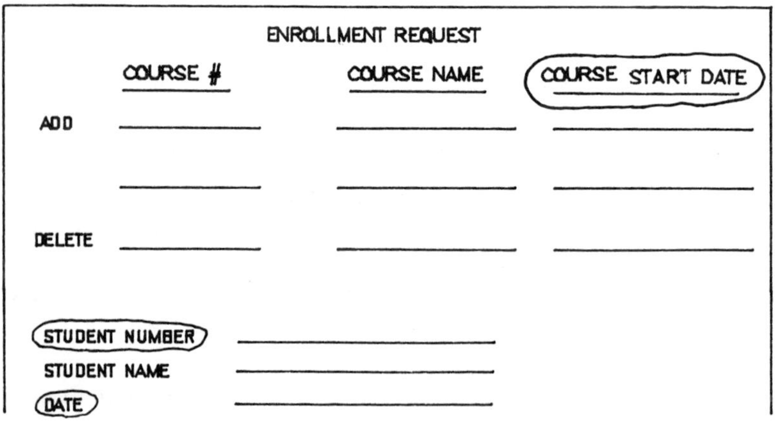

Figure 3.2.

From the Completions form, Figure 3.3, he gets several more: *instructor name*, *completion date*, and *no-show reason*.

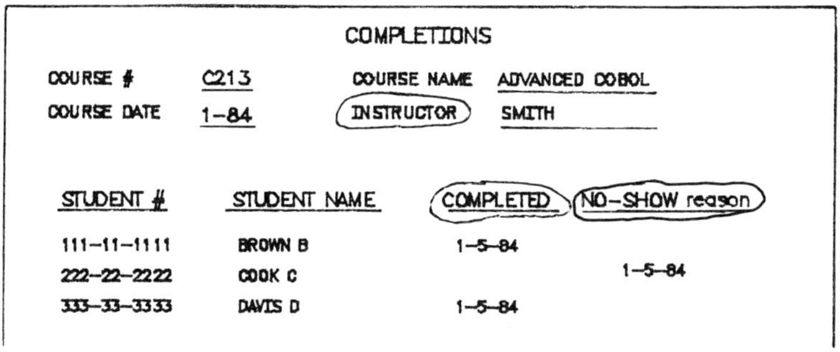

```
                          COMPLETIONS
       COURSE #    C213       COURSE NAME   ADVANCED COBOL
       COURSE DATE  1-84      INSTRUCTOR    SMITH


       STUDENT #      STUDENT NAME      COMPLETED    NO-SHOW reason
       111-11-1111    BROWN B           1-5-84
       222-22-2222    COOK C                         1-5-84
       333-33-3333    DAVIS D           1-5-84
```

Figure 3.3.

Alan has recalled from his interviews with the students and their managers that they did not like the current procedure for reporting no-shows and decided to print a no-show reason rather than just a date.

From the Course Schedule, Figure 3.4, he gets *hours, course end date, course time,* and *room location.* From the Completion report for January, Figure 3.5, he gets *group number* and *group manager name*. From the Needs form, Figure 3.6, he gets *student level, course level,* and *needed date.*

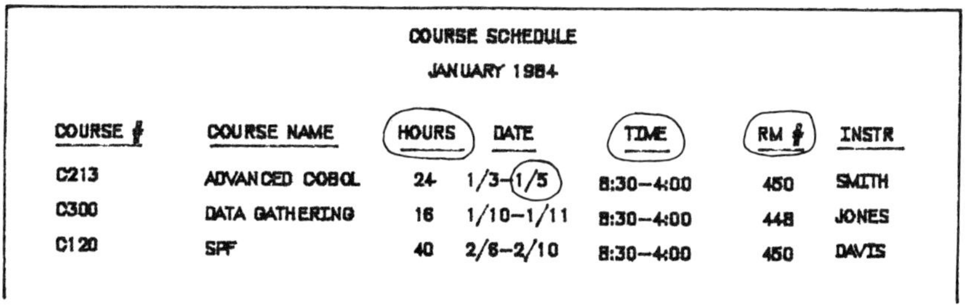

```
                         COURSE SCHEDULE
                          JANUARY 1984

   COURSE #    COURSE NAME      HOURS   DATE        TIME       RM #   INSTR
   C213        ADVANCED COBOL    24     1/3-1/5     8:30-4:00  450    SMITH
   C300        DATA GATHERING    16     1/10-1/11   8:30-4:00  448    JONES
   C120        SPF               40     2/6-2/10    8:30-4:00  450    DAVIS
```

Figure 3.4.

DP TRAINING REPORT FOR JANUARY 1984

(GROUP #:) 1   (GROUP MANAGER:) ADAMS

STUDENT NAME	COURSE #/NAME	COMPLETION DATE	NO-SHOW DATE
BROWN B	C213 ADVANCED COBOL	1-5-84	
	C300 DATA GATHERING		1-5-84
JONES J	C300 DATA GATHERING	1-11-84	

GROUP #: 2   GROUP MANAGER: WEEKS

STUDENT NAME	COURSE #/NAME	COMPLETION DATE	NO-SHOW DATE
COOK C	C213 ADVANCED COBOL		1-5-84
DAVIS D	C213 ADVANCED COBOL	1-5-84	
JONES J	C300 DATA GATHERING	1-11-84	
SMITH S	C300 DATA GATHERING	1-11-84	

Figure 3.5.

TRAINING NEEDS FORECAST 1984   JONES J   (STUDENT LEVEL: 2)
MANAGER NAME: WEEKS

COURSE #	COURSE NAME	(COURSE LEVEL)	COMPLETED	(NEEDED DATE)
C100	xxxxxxxxxx	1	1-15-83	
C110	xxxxxxxxxx	1	2-20-83	
C120	xxxxxxxxxx	1	4-10-83	
C200	xxxxxxxxxx	2	9-20-83	
C213	ADVANCED COBOL	2		1/84
C240	xxxxxxxxxx	2		3/84
C300	DATA GATHERING	3		1/84
C320	xxxxxxxxxx	3		4/84

Figure 3.6.

From the course description in the catalog, Figure 3.7, he gets *course description* and *course prerequisite*.

He notices that *hours* is called *credit hours* in the catalog and decides *credit hours* is more descriptive.

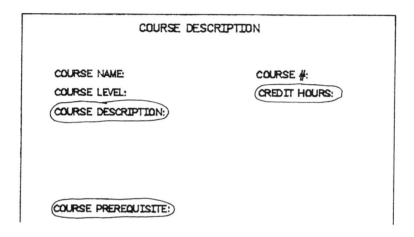

Figure 3.7.

He does not notice any new data items on the Training Statistics report, shown in Figure 3.8. Because both the number of students and the number of student days can be determined from other data items, he doesn't put them in the dictionary now.

```
              TRAINING STATISTICS

     Course #    # Students    Student Days    Instructor
     ────────    ──────────    ────────────    ──────────
       C213          2              6            Smith
       C300          2              4            Jones
                    ───            ───
                     4              10
```

Figure 3.8.

As Alan inspects the Forecasted Needs report, Figure 3.9, he notices that the only new data item is the *number of needs*, which can be determined from other data items.

```
                    FORECASTED NEEDS REPORT
                             1984
        COURSE #     COURSE NAME      QTR    # NEEDS

        C213         Advanced Cobol    1        3
        C300         Data Gathering    1        2
        C300         Data Gathering    2        1
        etc.
```

Figure 3.9.

Alan sends Terri a copy of his definitions so that she can review them for accuracy. He also asks her about the significance of the month on the Course Roster.

Questions

1. What is meant by business data? Data item?
2. How would Alan find the data items if this were an entirely new system that had not existed in a manual procedure?
3. What is meant by derived data? Do you want it in the data dictionary? Why?
4. What other things will Alan be putting into the data dictionary as he develops the system?

4

The Information Model

> *... it is the responsibility of the* true *DP professional to educate users about the power of the computer.*
> David King, *Current Practices in Software Development.*
> New York: YOURDON Press, 1984, p. 5.

Alan has received the corrected data dictionary definitions back from Terri. The following list is what she sent him.

Completion Date: The date an enrolled student completes all portions of a scheduled course.

Course Description: The narrative which describes the course content and objectives.

Course End Date: The last day of a scheduled course.

Course Level: The level of the intended course audience.

Course Name: The name used to describe a course provided by the data processing training group.

Course Number: A number which uniquely identifies a course provided by the data processing training group.

Course Prerequisite: The course number of the prerequisite course.

Course Start Date: The first day of a scheduled course.

Course Time: The beginning and ending time of the scheduled course. Applies to each day of a multiple-day course.

Credit Hours: The number of hours of training credit given for a course. One day is considered six hours.

Enrollment Date: The date a student is enrolled in a scheduled course.

Group Manager Name: The name of the manager of a work group.

Group Number: A number which uniquely identifies a work group.

Instructor Name: The name of a person assigned the responsibility for teaching the course.

Needed Date: The month and year a student anticipates needing to take training.

No-Show Reason: The reason why an enrolled student did not complete a course.

Room Location: The room number and building of the classroom.

Student Level: The job level of a student.

Student Name: The name of a student.

Student Number: The social security number of a student.

Alan reviews the documentation that he has produced so far. In addition to the data dictionary, his Requirements Definition document contains the problem statement, the user needs and objectives, the events to which the system must respond, and the context diagram. He refers to the system development standards manual to see what to do next.

The manual lists the next step as data analysis. As Alan reads the activities listed for data analysis, he realizes that this is where information modeling fits in. After several days of reviewing information modeling, he is just ready to start when his manager stops by to see how he is getting along.

THE INFORMATION MODEL 21

His manager approves of what Alan has completed for the requirements definition, but is somewhat puzzled about the data dictionary definitions. "Aren't you getting a little ahead of yourself here? I wouldn't think you'd need to know the data element definitions until you start coding. But I guess it won't hurt since you'll need it sooner or later anyway. What are you doing now?" Alan replies that he is working on the information model. "What is an information model?" his manager asks. "Will it help you do prototyping?"

Instead of answering, Alan asks, "Do you have a few minutes? I'll give you a quick overview to explain the basic principles.

Objects and relationships

"An information model consists of an object-relationship diagram, object definitions, and relationship definitions. Objects are the things that a business stores information about. An object can be something tangible, such as a person or place, or it can be something intangible such as a concept, agreement, or event. For example, an object can be an employee, a customer, a contract, or a ballgame. In an information model, you define the objects that your system records data about and how these objects are related to each other.

"Objects have properties, pieces of information which tell us something about the object. Each data item in the data dictionary is a piece of information which is attributed to one and only one object.

"A relationship is a connection between two or more objects. For instance, a vendor may sell many different products; a employee may have at most one spouse.

"Objects, relationships, and attributes could be thought of as nouns, verbs, and adjectives." See Figure 4.1.

Figure 4.1.

"EMPLOYEE and BUILDING are objects; 'works in' is the relationship between EMPLOYEE and BUILDING; *employee name* is an item of data that the business knows about EMPLOYEE, and *building name* is an item of data that the business knows about BUILDING. Of course, there may be many other data that the business needs to know about EMPLOYEE and BUILDING.

"There are several different kinds of objects. If you find as you are attributing data to an object that some of the items actually describe another item, pull that item out and make it an object in its own right. This is called a *characteristic object.*" See Figure 4.2.

```
         ┌──────────┐
         │ EMPLOYEE │
         └──────────┘

         employee name
         employee address
         employee age
         employee spouse
         employee spouse address
         employee spouse age
```

Figure 4.2.

"*Employee spouse address* and *employee spouse age* actually describe *employee spouse* rather than EMPLOYEE, so *employee spouse* is pulled out and becomes a separate object." See Figure 4.3.

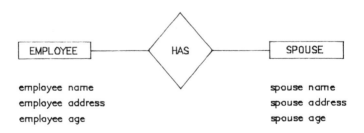

Figure 4.3.

"If you find that a group of items seem to describe a relationship or an object only in its relationship to some other object, make that relationship an object." See Figures 4.4 and 4.5.

THE INFORMATION MODEL 23

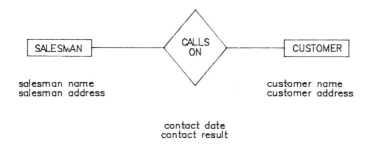

Figure 4.4.

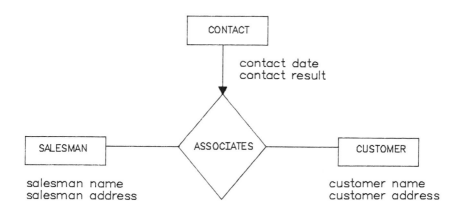

Figure 4.5.

"A relationship which becomes an object is called an *associative object* and is shown by an arrow pointing to the relationship. You will often find that formal documents are associative objects.

"You may also find that some objects have data that describe the object only when it assumes a particular role, for example, ACTIVE CUSTOMER, POTENTIAL CUSTOMER, and INACTIVE CUSTOMER. These are referred to as *subtypes* of CUSTOMER. Other objects may have relationships either to the subtypes or to the supertype (CUSTOMER). When attributing data items to them, attribute data that describe them all to the supertype and only the unique data to each subtype." See Figure 4.6.

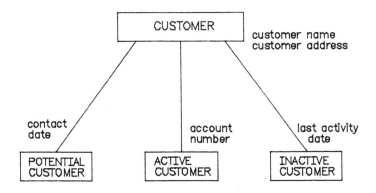

Figure 4.6.

"That pretty much covers the principles of information modeling. Sounds simple, doesn't it?" Alan's manager acknowledges that an information model makes sense, but wants to know what benefit it will be to Alan's project. Alan explains that there are several reasons for producing an information model.

"One benefit is that it will reduce redundancy in the data stores, regardless of how the data is organized for storage on the tapes or disks. Better file design means a more stable system; a more stable system means less and easier maintenance. Another benefit is that the data will be grouped according to the way the application views it. As a result, data flow diagrams will be much quicker and easier to develop.

"Well," his manager replies, "go ahead and try it. I'm for anything that will help you do structured analysis more quickly. Structured analysis sure seems to take a long time on most projects. Frankly, I've never been convinced that it was worth it."

Identifying objects

"Whew," Alan thinks. "I'm glad my explanation made some sense to him. I just hope I can actually do an information model for my application. It may sound simple, but I know it's not quite so easy to do as it sounds. I'd better start trying to identify the objects."

To identify the objects, Alan reads through his data dictionary definitions and quickly picks out some of the main objects. These are COURSE, SCHEDULED COURSE, STUDENT, GROUP, INSTRUCTOR, NEEDED COURSE, and ROOM. Noticing that there seem to be three types of courses — NEEDED COURSE, COURSE, and SCHEDULED COURSE — he suspects that SCHEDULED COURSE and NEEDED COURSE are subtypes of COURSE. He also notices the use of STUDENT and ENROLLED STUDENT and wonders if a student who needs a course would be a POTENTIAL STUDENT. Probably ENROLLED STUDENT and POTENTIAL STUDENT are subtypes of STUDENT. He defines each object as follows:

Object: Course
A course offered by the data processing training group.

Object: Enrolled Student
A student who is enrolled in a scheduled course.

Object: Group
The work group in which a potential student works.

Object: Instructor
A person qualified to teach a course.

Object: Needed Course
A course for which a potential student has expressed a need.

Object: Potential Student
Someone who is eligible to take a course.

Object: Room
A meeting place for an instructor and a student.

Object: Scheduled Course
A course which has been assigned dates, room, and instructor.

Object: Student
Someone who needs a course or enrolls in a scheduled course.

After drawing an object-relationship diagram and naming the relationships, Alan needs to review his work with Terri to be sure that he has correctly understood the objects and their relationships. Together they will attribute data items to the objects. His object-relationship diagram looks like Figure 4.7.

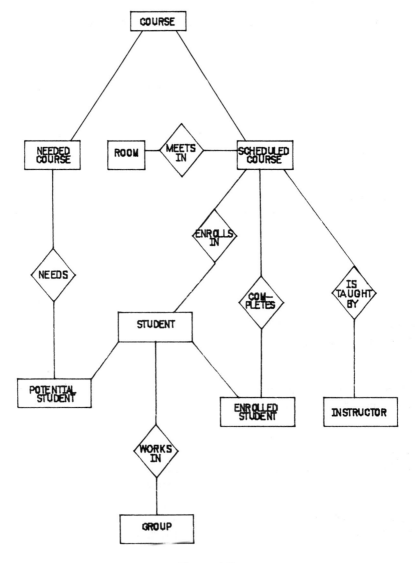

Figure 4.7.

Alan calls Terri, but she won't have time to meet with him until next week.

"No wonder system development takes so long!" Alan thinks. "Seems like I spend half my time waiting to get the information I need. And I have a cooperative user! Oh, well, I'd better not just sit here or my manager will think I don't know how to do this. I'll work on my own while I'm waiting to talk to Terri."

Attributing Data Items to Objects

He starts going through the data dictionary list and tries to attribute each item to an object. If a new object is discovered, he'll add it to the diagram and define it.

Completion Date: The date an enrolled student completes all portions of a scheduled course.

The definition mentions two objects — ENROLLED STUDENT and SCHEDULED COURSE — and describes their relationship, drawn in Figure 4.8.

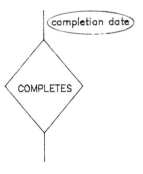

Figure 4.8.

Course Description: The narrative which describes the course content and objectives.

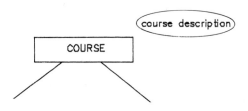

Figure 4.9.

Course End Date: The last day of a scheduled course.

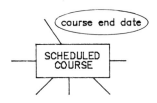

Figure 4.10.

Course Level: The level of the intended course audience.

Course Name: The name used to describe a course provided by the data processing training group.

Course Number: A number which uniquely identifies a course provided by the data processing training group.

Course Prerequisite: The course number of the prerequisite course.

THE INFORMATION MODEL 29

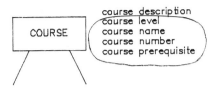

Figure 4.11.

Course Start Date: The first day of a scheduled course.

Course Time: The beginning and ending time of the scheduled course. Applies to each day of a multiple-day course.

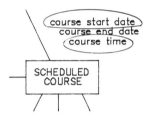

Figure 4.12.

Credit Hours: The number of hours of training credit given for a course. One day is considered six hours.

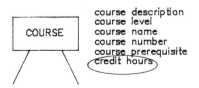

Figure 4.13.

Enrollment Date: The date a student is enrolled in a scheduled course.

30 THE STRUCTURED SYSTEMS LIFE CYCLE: A CASE STUDY

Alan starts to attribute *enrollment date* to STUDENT, but realizes that it also has something to do with courses. Then he starts to attribute the entry to SCHEDULED COURSE, but realizes that it also has something to do with students. He rereads the definition which defines it as the date a student (object) enrolls in a scheduled course (object). *Enrollment date* describes the relationship between the two objects.

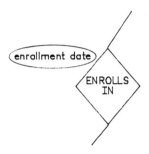

Figure 4.14.

Group Manager Name: The name of the manager of a work group.

Group Number: A number which uniquely identifies a work group.

Figure 4.15.

Instructor Name: The name of a person assigned the responsibility for teaching the course.

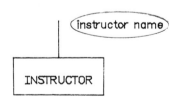

Figure 4.16.

Needed Date: The month and year a student anticipates needing to take training.

Figure 4.17.

No-show Reason: The reason why an enrolled student did not complete a course.

No-show reason describes STUDENT. Oops! No-show for what? No-show for a course in which the student had enrolled. *No-show reason* must be an attribute of the relationship between a SCHEDULED COURSE and a STUDENT.

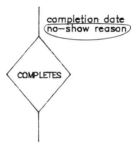

Figure 4.18.

Room Location: The room number and building of the classroom.

Figure 4.19.

Student Level: The job level of a student.

Student Name: The name of a student.

Student Number: The social security number of a student.

Figure 4.20.

Alan's first-cut information model is shown in Figure 4.21.

Now the model needs to be reviewed and evaluated, and he cannot do that until he meets with Terri.

THE INFORMATION MODEL 33

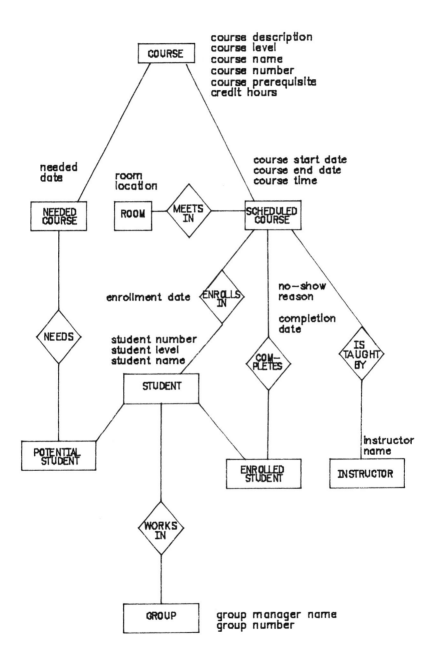

Figure 4.21.

Questions

1. What is an object?
2. What is an attribute?
3. Define the different types of objects and explain how you know which type an object is.
4. What is another name for an object-relationship diagram?
5. If Alan finds that a subtype object has no attributes that uniquely apply to it, what should he do? What if the only attribute of a subtype object is its number or name?
6. Was Alan smart to continue work on the information model before he could talk to Terri about it?
7. How could Alan have avoided wasting time waiting to talk to Terri?
8. Do you know of any techniques (other than prototyping) for speeding up requirements definition?

References

Blash, Greta. "Information Modeling: A Data Modeling Strategy," Guide 59, November 1984.

Date, Chris. *Introduction to Database Systems*, 3rd edition. Reading: Addison-Wesley, 1981.

Flavin, Matt. *Fundamental Concepts of Information Modeling.* New York: YOURDON Press, 1981.

5

The Refined Information Model

To apply mathematical dependency analysis without understanding and articulating the underlying policy can be compared to the experience of shooting in the dark.
Matt Flavin, *Fundamental Concepts of Information Modeling.*
New York: YOURDON Press, 1981, p. 2.

Alan has just entered Terri's office. He has brought his information model and the data dictionary definitions.

"Alan, it's nice to see you again. How are you coming along with the project?"

"Quite well, actually. In fact, things are going so smoothly that I keep wondering what I'm doing wrong."

"Or it could just be that things go better when proper procedures are followed," Terri suggests.

"I hope that's it."

"What is it that you need to review with me?" Terri asks.

"Are you familiar with information modeling?"

"I've heard the term, but I've never actually seen an information model."

"I built one for our project because when I learned about it in college I saw how useful it can be," Alan begins.

36   THE STRUCTURED SYSTEMS LIFE CYCLE: A CASE STUDY

Reviewing the first-cut information model

"I do need to review it with you, though, because I made some assumptions that may not be correct. Let's start with the easy parts first. You store information about STUDENT — *student number, student level,* and *student name.* Students work in a GROUP. You record the *group number* and the *group manager name.* Is that correct?"

"Yes. Go on."

"I see two possible subtypes for STUDENT — POTENTIAL STUDENT and ENROLLED STUDENT. However, I could not find where you recorded any unique information about them."

"That's probably true," Terri says.

"Another object is COURSE. You record *course description, course level, course name, course number, course prerequisite,* and *credit hours.*

"There are two possible subtypes of COURSE — NEEDED COURSE and SCHEDULED COURSE. For SCHEDULED COURSE, you record *course start* and *end dates* and *course time.* As I understand it, a POTENTIAL STUDENT expresses a need for a course to be scheduled sometime in the future. The only data item I could find that you recorded about NEEDED COURSE is the *needed date.* Is that correct?"

"Yes, I only use that information to help me with my scheduling."

"Okay. Let's go on. A SCHEDULED COURSE meets in a ROOM and is taught by an INSTRUCTOR. A STUDENT enrolls in a SCHEDULED COURSE. An ENROLLED STUDENT completes a SCHEDULED COURSE. Is everything I've said true?"

"Yes. Why do you show some data items beside the relationships rather than beside the objects?" Terri asks.

"Well, *enrollment date* has no meaning by itself because it is the date that a STUDENT enrolls in a SCHEDULED COURSE. The same applies to *completion date* and *no-show reason.* They only have meaning in conjunction with a particular STUDENT and a particular SCHEDULED COURSE."

"What do you mean by SCHEDULED COURSE?"

"A SCHEDULED COURSE is an occurrence of a course that's in the curriculum. We could have called it a class, but I liked the consistency of using the word course. I can call it class if you'd prefer."

"No, SCHEDULED COURSE is fine," Terri says.

"Now, I want to be sure that I'm planning to store all the information you're going to need. Do you want to know anything about ROOM other than the location?"

"No, not right now. What will happen if later on I want to keep track of what type of facilities the rooms have or how many people they'll hold?"

"It shouldn't be much problem to add that later if I do my job right," Alan answers. "What about INSTRUCTOR? Any information you would like to have other than the name?"

"No, not now. Some day I might like to know what each instructor's skills are, but I don't even have that information available to me right now."

Eliminating objects

"In that case, let's simplify our information model by eliminating ROOM and INSTRUCTOR as objects and make *room location* and *instructor name* attributes of SCHEDULED COURSE."

"I don't quite understand your reasoning. What if later I want to store the other information I mentioned?"

"No problem. I'll just make ROOM and INSTRUCTOR objects again. Can you think of anything different you'd like to store about either POTENTIAL STUDENT or ENROLLED STUDENT that doesn't apply to STUDENT in general?"

"No, I don't think I make any distinction," Terri says. "A student is a student as far as I'm concerned."

"In that case, I'm going to eliminate POTENTIAL STUDENT and ENROLLED STUDENT as objects, and our model will look like this diagram." See Figure 5.1.

"Looks a little less cluttered, doesn't it?" Alan remarks.

"You know, this thing actually makes some sense to me. Anything else you need to know?"

38 THE STRUCTURED SYSTEMS LIFE CYCLE: A CASE STUDY

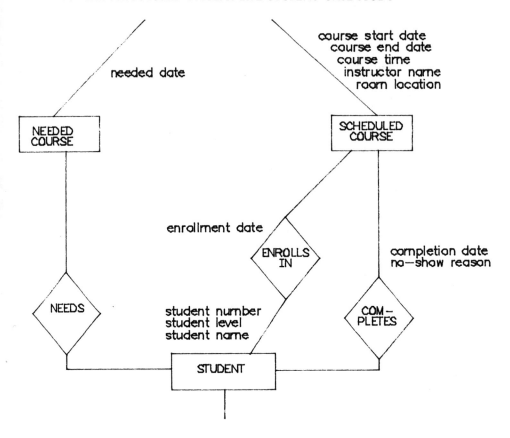

Figure 5.1.

Eliminating multiple relationships

"I'm not sure yet. First, let me fix something on the diagram. When we removed ENROLLED STUDENT, that left us with two relationships between SCHEDULED COURSE and STUDENT. Let's just show one relationship." See Figure 5.2.

Adding an associative object

"Whenever we end up with a relationship that has attributes, we create what we call an associative object," Alan explains. "That is, an object which may exist solely for the purpose of associating two other objects. What is it that associates STUDENT and SCHEDULED COURSE?"

"I would say it would be a registration," Terri says.

"Fine, that's what we'll call it. Our model now looks like this." See Figure 5.3.

THE REFINED INFORMATION MODEL 39

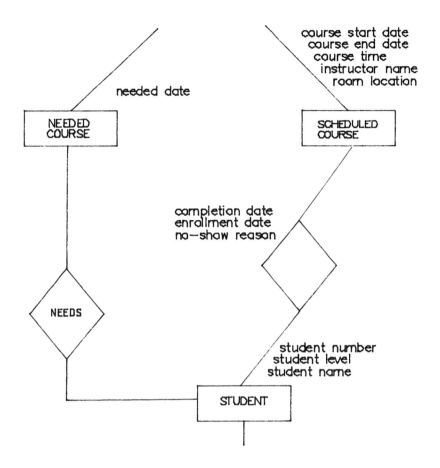

Figure 5.2.

Attributing derived data

"Does that seem to correctly reflect the relationships among the objects?" Alan asks.

"Yes, but something about NEEDED COURSE bothers me."

"What's that?"

"I'm not quite sure. If the only information we are going to store is *needed date*, why don't we just attach it to some other object like we did with room location and instructor name?"

40 THE STRUCTURED SYSTEMS LIFE CYCLE: A CASE STUDY

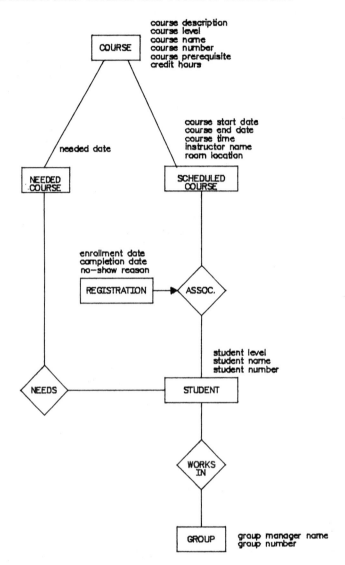

Figure 5.3.

"Hmmm. I guess we could attach it to COURSE, but something about this is bothering me, too. Maybe there's something we aren't thinking of. I just noticed that we don't have anything to uniquely identify one request from another. How do you tell one request for a course from another?"

"I don't," Terri says. "All I'm interested in is the total number of requests for a given course for a particular month. One of the reports I sent you must have had a column titled 'Number of Needs,' didn't it?"

"Yes, but we don't normally show data on the information model that can be derived from other data. And I figured we could just add up the requests."

"But that means that you would have to record a count or something, wouldn't you?" Terri asks.

"Yes, I guess I would. Let's just go ahead and say we'll store the number of needs, even though it could be derived from other data. I think that will solve our problem. Take a look at this diagram now. Is there anything else you notice that doesn't seem quite right?"

The refined information model

"I was just looking at REGISTRATION. Students fail to complete courses for reasons other than just not showing up. I think we should call it *withdrawal reason* rather than *no-show reason.*"

"No problem. Does it look okay now?" See Figure 5.4.

"Looks good to me," Terri says, "but wouldn't the diagram become crowded if we had many more data items we wanted to store?"

"Good point. In fact, we may not leave the data items on the diagram in our finished documentation. They're part of the object definitions anyway. It just makes it a lot easier to review when we don't have to handle several sheets of paper."

"That makes sense. Are we done with our information model now?" Terri asks.

"Nearly. I just need to write a statement which defines each of the relationships on our diagram. I'll need that information later on to design the files. I assume a student can be registered for more than one scheduled class?"

"Yes."

"Are there students who have never enrolled in a course?"

"Yes."

"Is it possible to have a scheduled course for which there are no students registered?"

"I suppose it's possible, but in that case I would just cancel the class."

42 THE STRUCTURED SYSTEMS LIFE CYCLE: A CASE STUDY

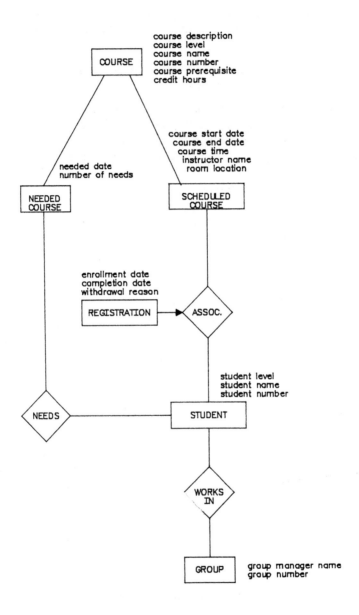

Figure 5.4.

"Let me show you the format we use to define this relationship. We'll first state the relationship between the associative object and the other objects. Then we'll state the relationship from the point of view of each of these other objects."

Relationship: Associates

For each registration, there is a scheduled course and a student such that these objects are associated when a student enrolls in a scheduled course.

Reverse Relationship:

1. For each student, there is none, one, or many registrations.
2. For each scheduled course, there is none, one, or many registrations.

"Does this correctly describe the relationship among SCHEDULED COURSE, STUDENT, and REGISTRATION?" Alan asks Terri.

"Seems okay to me."

"Can a student work in more than one group at the same time?"

"Not so far as the official records go, and that's all I care about."

"Does the following correctly describe the relationship between STUDENT and GROUP?"

Relationship: Works In

For each student, there is a group such that each student works in one group.

Reverse Relationship:

Each group may contain one or many students

"I think so," Terri answers.

"What about this description for the relationship between NEEDED COURSE and STUDENT?"

Relationship: Needs

For each needed course, there is a student such that each needed course is needed by one or many students.

Reverse Relationship:

Each student has none, one, or many needs.

"Yes," Terri says. "Now that I understand what you're trying to accomplish by doing the information model, I've been wondering why you didn't just ask me what I needed to know about each of the objects instead of going through the documents to find out."

"I thought about doing that, but was afraid we'd miss something. I could have used the documents just to double-check, though. If this system had not been an existing manual procedure, I wouldn't have had any existing documents to analyze. In that case, the only approach would have been to interview the customer."

"What do you do next?"

"The next thing I'll be doing is making sure that I understand all of the processing that the system will need to do. While I'm here, could I schedule a few hours of your time to talk about that?"

"Sure, how about two weeks from today?"

"That will be great," Alan says. "See you then. And thanks again."

Alan now can add the object-relationship diagram, object definitions, and relationship definitions to the Requirements Definition document.

Questions

1. If Alan were more experienced, what question would he have asked Terri when she kept talking about the room and instructor data?

2. Is it possible for an object to have different attributes depending on the application?

3. What determines whether or not an object is included in the information model for a particular system?

4. What test can be used to tell if an object is associative?

5. Alan eliminated the objects ROOM and INSTRUCTOR because he had only one data item attributed to each. Under what condition would he *not* have eliminated an object that had only one attribute?

6. What do we know about the relationship between objects and data stores?

7. Why do we attribute a data item to only one object?

8. Why do we define the relationships so precisely?

9. What approaches can be used to determine the data items which describe the objects in a system.

References

Date, Chris. *Introduction to Database Systems.* Reading: Addison-Wesley, 1981.

Flavin, Matt. *Fundamental Concepts of Information Modeling.* New York: YOURDON Press, 1984.

6

The Process Model

> *I will contend that conceptual integrity is* the *most important consideration in system design . . . . Conceptual integrity in turn dictates that the design must proceed from one mind, or from a very small number of agreeing resonant minds.*
> Frederick P. Brooks, Jr., *The Mythical Man Month.*
> Reading: Addison-Wesley Publishing Co., 1975, pp. 42 and 44.

Alan has been thinking about the best way to determine the processsing requirements of the application that he is analyzing.

In order to draw the data flow diagrams, he has to know what happens to each input, and he does not understand the training system well enough to be sure how everything works. He could wait until he talks with Terri, ask her questions that he has prepared, come back to his desk, draw the data flow diagrams, and then review them with her.

Another approach would be to draw the data flow diagrams as accurately as he can, prepare a list of questions as he does it, and talk to her just once about it. He does have a fairly good idea how the system must work, probably enough to draw a first-cut data flow diagram. Terri is used to reading data flow diagrams, so she is not likely to be intimidated if he brings them to their next appointment. He decides to draw them the best he can, make a list of questions, and hope that approach will save a little time.

Diagramming events

"I already know what my data stores are because they're the same as the objects in the information model," Alan thinks. "Let's see: Curriculum Courses, Needed Courses, Scheduled Courses, Registrations, Groups, and Students. I know what the system inputs and outputs are. I should be all ready to start.

"Oh, yes. I need to associate each system input with the event that triggers it. Terri already told me the events which this system must respond to. Where did I file them? Here they are."

Event	System Input	Expected Response
Job skill requirements change.	Skill Requirements	Curriculum is designed.
Student foresees training need.	Forecasted Need	Curriculum courses are scheduled.
Student decides to take a course.	Enrollment	Student is trained.
Personnel change occurs.	Personnel Information	Student and/or group information is changed.

Alan is going to draw one bubble for each event and then consolidate the diagrams. This first-cut diagram will not be perfect, but it will show the input and output data flows required to process each event. Even with just one bubble per event, there will be too many processes for the final Level 1 data flow diagram if the system is very large. But it is a quick way to start.

Another advantage to event partitioning is that events can be assigned to different team members to work on concurrently. This can be especially helpful if the application being analyzed is large. Usually, the completed pieces will fit together without too much rework.

Event 1

Alan starts by diagramming what occurs when skill requirements change for a job. Referring to the notes from his first interview with Terri, he recalls that she had told him that when skill requirements change, she may have to redesign the curriculum. She also had said that her department produces a Course Catalog, which contains a

48 THE STRUCTURED SYSTEMS LIFE CYCLE: A CASE STUDY

description of all the courses in the curriculum. Alan produces Figure 6.1.

Event: Job skill requirements change.

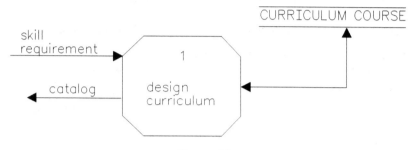

Figure 6.1.

Event 2

Now Alan considers what happens when students foresee the need for a course. Again checking his interview notes, he reads that Terri had told him that she was concerned only with the total number of requests for a particular course in any future month. She uses this information to help decide when to schedule courses. The course Schedule is published each month. He draws a data flow diagram for these processes in Figure 6.2.

Event: Student foresees training need.

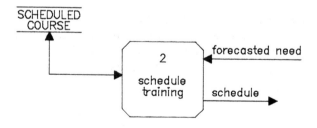

Figure 6.2.

THE PROCESS MODEL 49

Alan has been diagramming the events in the order in which he has them listed, but he could have started with the easiest ones, or with the ones he understands the best, or in the order in which they actually occur. These data flow diagrams eventually will be consolidated, evaluated, and refined anyway.

Event 3

Next, Alan diagrams what happens when a student decides to enroll in a scheduled course. Terri had mentioned producing class Rosters. Alan knows that the students and their managers get the Rosters, but he guesses that the instructors must get them also so that they will know who is supposed to be in the class. Alan also notices from his notes that Terri had commented that there is a Completion form used only within the training area. He figures that it must be used to record who completed the course and who was a no-show. Alan draws a data flow diagram for these functions in Figure 6.3.

Event: Student decides to take a course.

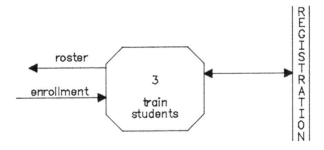

Figure 6.3.

Event 4

The last event on Alan's list is a personnel change. When Personnel Information changes, either the Student or the Group data store may need to be updated. Alan shows these in Figure 6.4.

Event: Personnel change occurs.

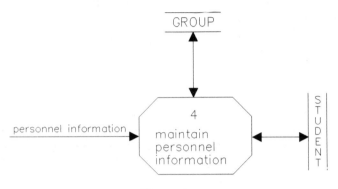

Figure 6.4.

Drawing a consolidated event diagram

Alan has diagrammed all the events and started to consolidate the four data flow diagrams into one. He has just crumpled up his sixth piece of paper and is tossing it into his wastebasket when he looks up to see his manager.

"Things not going well today, Alan?"

"Oh, nothing serious. Just trying to consolidate these event diagrams into one nice looking data flow diagram to use for the walkthrough with Terri," Alan replies.

"Why are you consolidating them?" his manager asks. "I thought one of the big arguments for data flow diagrams is that they break the system down into small diagrams that can be analyzed one at a time."

"That's true," Alan says, "but it's also helpful to be able to look at the whole system at once. This is just a working document, and it doesn't need to be maintained once the system is operational. I want it to take to my next meeting with Terri so that we can see if I have correctly understood how the application works. We'll look for missing data flows between processes, redundant processes, illogical data flows, and so on."

"Why are you drawing it by hand?" his manager asks. Seeing Alan's puzzled look, he continues, "Hasn't anyone told you that there are software packages now that draw data flow diagrams, structure charts, entity-relationship diagrams, and about anything else you could

THE PROCESS MODEL 51

want? The development center has one. They'll show you how to use it."

"Thanks. I'll do that."

Alan spends half of a day learning how to use the software package and produces his consolidated first-cut data flow diagram in Figure 6.5. Data flows connecting the event diagrams will be added as the model is refined.

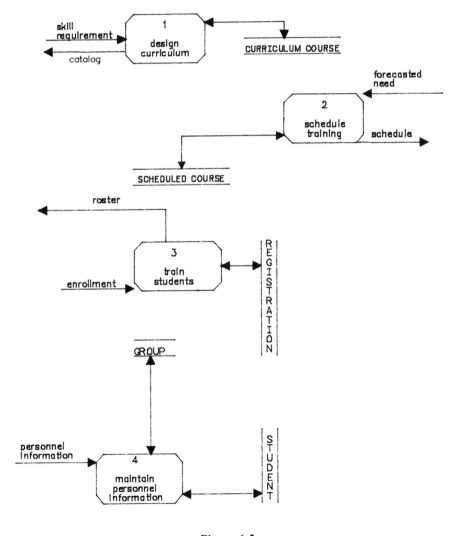

Figure 6.5.

52 THE STRUCTURED SYSTEMS LIFE CYCLE: A CASE STUDY

Alan refers to the context diagram, shown in Figure 6.6, to see if he has included all the system inputs and outputs.

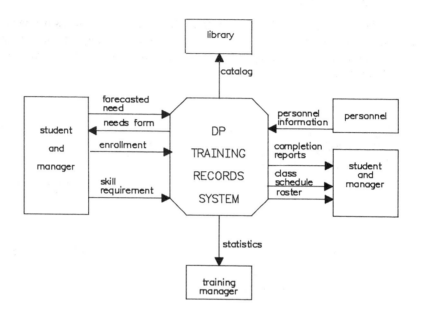

Figure 6.6.

He notices that he has missed Completion reports, Statistics, and the Needs form. Alan is not sure what causes them to be produced, but he can tell from the data on them what data stores are used. He adds them to the data flow diagram.

He also surmises that the Catalog probably is used in the process of scheduling training, so he adds that data flow and gets Figure 6.7.

THE PROCESS MODEL 53

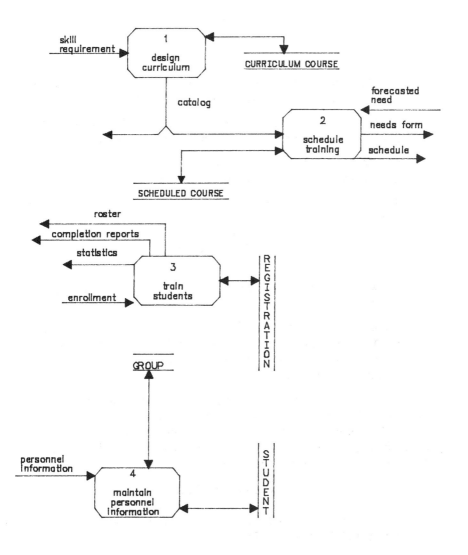

Figure 6.7.

"Hmmm," he says. "Not too bad for my first try! Now I just need to make a list of some questions to ask Terri so that I can refine the model and write my process descriptions."

Questions

1. Do you detect anything that Alan omitted?
2. Do you think the level of detail on Alan's data flow diagram is appropriate for a Level 1 Data Flow Diagram? Why?
3. Alan has shown some processes that may not be mechanized or that are a mix of manual and mechanized. Should he?
4. Can an event have more than one system input associated with it? Can an event have system outputs associated with it?
5. The data store Needed Course does not appear on the diagram. Why?

Exercise

Draw a context diagram to depict the following transaction processing system.

Transactions are large or small, red or blue, and round, square, or oblong. All transactions must be edited for size, shape, and color. An error report showing which errors were found in each rejected transaction is returned to the user.

An amount is calculated for large transactions. This amount is the quantity in the transaction multiplied by a calculation factor stored in a table. All transactions, including the calculated amount, are written on a file. Red square and blue round transactions are printed on an exception report, which goes back to the user.

References

DeMarco, Tom. *Structured Analysis and System Specification.* New York: YOURDON Press, 1978.

Palmer, J. and S. McMenamin. *Essential Systems Analysis.* New York: YOURDON Press, 1984.

Ward, Paul T. *Systems Development Without Pain.* New York: YOURDON Press, 1984.

7

The Refined Process Model

The principal lesson of the first 25 years of data processing is that software development is harder to manage and control than it appeared to be at the outset. Without a clean and compelling design, a large application system soon becomes a jumble of confusion and frustration.

Harlan D. Mills, *Software Productivity*.
Boston: Little, Brown and Co., 1983, p. 237.

With the first-cut data flow diagram spread out on her desk, Alan is showing Terri how he analyzed the processing required for each system input and explaining some of his assumptions. Terri verifies his assumption that she uses the course Catalog when she determines which courses to schedule. He also had assumed correctly that the instructors receive a copy of the course Roster and that they record completion information on the Completion forms.

Walking through the event diagram

Alan asks Terri if she will look at the data flow diagram, Figure 7.1, and see if he has missed anything, reminding her that at this phase he is concerned only with the data and processes that are essential to the system. Exactly how that processing will be done and who or what will process the data will be determined later.

56 THE STRUCTURED SYSTEMS LIFE CYCLE: A CASE STUDY

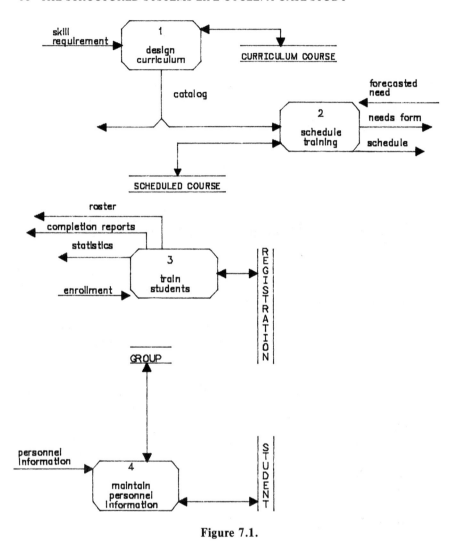

Figure 7.1.

"How will you make sure that you don't schedule courses that are not currently in the curriculum?" Terri asks.

"I could check the Curriculum Course data store before adding them."

"You also need to be sure you don't let students enroll in courses that haven't been scheduled."

"Okay." Alan agrees. "Also, I need to be sure that we don't add students for which no group has been set up."

"Right. I notice you show enrollments, but what about withdrawals? Were you including them with enrollments?"

THE REFINED PROCESS MODEL 57

"No, I wasn't. Isn't a withdrawal just the deletion of the enrollment?"

"Not always. If it's within ten days of the class, I require a withdrawal reason."

"Oh. I forgot about that. I'll add it."

"Do you have the information model with you?" Terri asks. "I don't remember what data items are in each of the data stores."

"Right here. Anything in particular you're wondering about?"

"Yes. Where are you getting the names for the Completion reports? Are they in the Registration data store?"

"No, they aren't. I'm afraid I forgot to check every data item on the reports. I'll need to check all of them. Thanks for catching that."

Alan's revised data flow diagram looks like Figure 7.2.

"Does that diagram seem to correctly represent your application now?" Alan asks.

"Yes, I think so. When will you have the Requirements Definition document completed?"

"I have nearly everything I need for the internal design of the system. I just have to write the narrative for each process on the diagram and define the information requirements for the system inputs and outputs. For the external design, though, we still need to design the screen and report layouts."

"That sounds like a lot of work still to be done. Is it going to take a long time?"

"The part that usually takes the longest is deciding how the screens should look. I'm going to try something new, however," Alan explains. "I'm going to use a fourth generation language to quickly prototype some sample screen interactions. I'll show them to you and we can make changes until they look the way you want them. Then I'll make a hard copy of them to put into the Requirements Definition document."

"That's a good idea. It's so hard to tell from screen layouts how they'll really look on the screen. Can we do that for the reports also?"

"Sure, I'll print several different sample layouts for each report to give you some ideas. Then you can decide exactly what formats you like before I assemble the Requirements Definition document."

58 THE STRUCTURED SYSTEMS LIFE CYCLE: A CASE STUDY

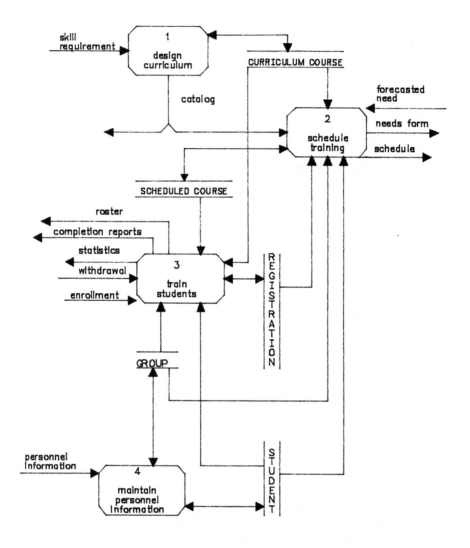

Figure 7.2.

"I'll be waiting to hear from you when you have the prototype ready."

"It won't be too long," Alan says. "I'll also probably need to ask you a few questions about the volume and frequency of the system inputs and outputs, but I can ask you that over the phone. Thanks so much for taking the time to go over these diagrams with me."

Describing the major processes

Returning to his desk, Alan starts describing the processes on the data flow diagram. He has been taught to write process notes for the higher level data flow diagrams. Unless notes are made about what will be in each major process, the analyst may have forgotten by the time the bottom level is reached, especially if the system is very large or if the work is interrupted for any length of time.

The first major process that Alan describes is "design curriculum," using Figure 7.3.

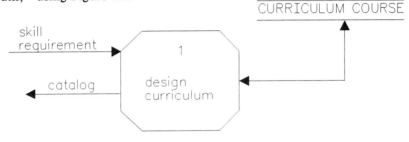

Figure 7.3.

Process Name: Design Curriculum
Process Number: 1
Process Notes:

When Skill Requirements change, the curriculum is analyzed to determine if courses should be added or dropped from the curriculum or if courses should be modified. After these determinations are made, Curriculum Course records are updated and revisions to the course Catalog are printed.

Alan next describes "schedule training," using Figure 7.4.

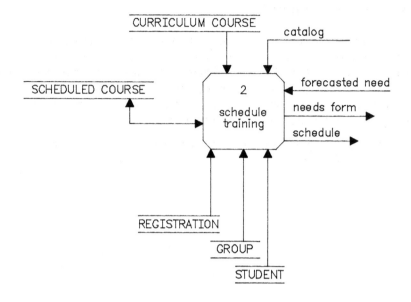

Figure 7.4.

Process Name: Schedule Training
Process Number: 2
Process Notes:

A Needs form is printed for each student. Each form contains a list of all the Curriculum Courses, showing which ones have been completed and giving a place to show when the student foresees a need to take any uncompleted courses.

When Forecasted Needs are received from the students, the total number of students expressing a need for each course is printed by NEEDED DATE.

The courses are scheduled according to the greatest need and the available room and instructor resources. The course Catalog is used to determine the course length and instructor requirements for the course. Scheduled Course records are updated and a revised class Schedule is printed.

The next process to be described, with Figure 7.5, is "train students."

THE REFINED PROCESS MODEL 61

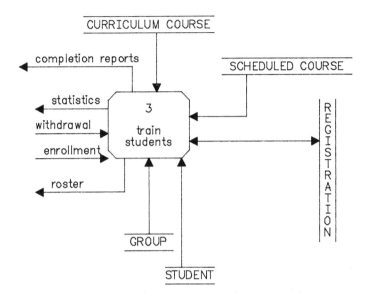

Figure 7.5.

Process Name: Train Students
Process Number: 3
Process Notes:

Students enroll in and withdraw from Scheduled Courses. Course Rosters are printed that list the students who are registered in each class. The instructor teaches the class and records which registered students complete the class and which do not.

Completion Reports and Training Statistics are printed.

The last major process to be described is "maintain personnel information," using Figure 7.6.

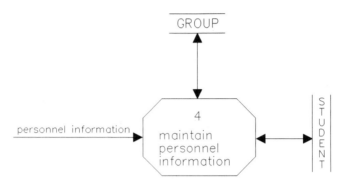

Figure 7.6.

Process Name: Maintain Personnel Information
Process Number: 4
Process Notes:

When personnel changes occur, both Group records and Student records are updated.

Defining information requirements

Alan next defines the information requirements for the system inputs and outputs. He called Terri to get the information about volume, frequency, retention, and security.

Completion Reports has an output volume of eight pages per month, and the major data elements are *group manager name, student name, course number, course name, course start date, completion date,* and *withdrawal reason.* Completion Reports are not retained, and there are no special security requirements.

Catalog has an output volume of one to two hundred pages, as required. Major data elements are *course number, course name, course level, course description, course prerequisite,* and *credit hours.* It is retained one cycle, and there are no security requirements.

Enrollment has an input volume of fifty to one hundred per month, and the major data elements are *course number, course start date, student number,* and *enrollment date.*

Forecasted Need has an input volume of two hundred to five hundred annually, and the major data elements are *course number* and *needed date*.

Needs form has an output volume of three hundred pages semiannually, and the major data elements are *group manager name, student name, course number, course name,* and *completion date*. It is retained for one cycle, and there are no special security requirements.

Personnel Information has an input volume of three hundred per month, and the major data elements are *student number, student name, student level, group number,* and *group manager name*.

Roster has an output volume of two to four pages monthly, and the major data elements are *course number, course name, course start date,* and *student name*. Rosters are retained three months and there are no special security requirements.

Schedule has an output volume of two pages monthly, and the major data elements are *course number, course name, course start date, course end date, course time, instructor name,* and *room location*. Schedules are retained for three months, and there are no special security requirements.

Skill Requirement has an input volume of two to five per year, and the major data elements are not applicable because this input is used only in a manual function.

Statistics has an output volume of two pages quarterly, and the major data elements are *course number, course name, course start date, number of students, number of students days, instructor name,* and *room location*. Statistics are retained for one year, and there are no special security requirements.

Withdrawal has an input volume of ten to twenty per month. Major data elements are *course number, course start date, student number,* and *withdrawal reason*.

Alan is checking the context diagram to be sure that he has defined the information requirements for all the system inputs and outputs when his manager calls him. "Alan, could you come in here a minute? I need to discuss something with you."

Changing plans

When Alan reaches his manager's office, his manager tells him, "I've just had a call from Terri's manager. As part of our new cost-cutting effort, they're going to have to start calculating the cost of training and bill each work group quarterly."

"That shouldn't be too difficult to add to the specifications," Alan replies.

"I realize that, but the real problem is that they'll have to be able to start capturing that information in two months. Terri says she can't do it unless she has this new system you're designing for her. That means that instead of having the luxury of no deadline for your first project, you have a very tight schedule.

"I've talked to some of the other managers, though, and I can borrow two additional people to help you. One of them has about twenty years experience. He's a super COBOL programmer and can debug better than anyone I know.

"The only thing that might be a problem is that he hasn't had any training on these new techniques. The way it looks, you won't have time to do much design anyway, and he could start coding and testing right now. We probably won't have time to do prototyping either, and I was really hoping we could tell people we had used it on this project. Oh, well, that's the way it goes.

"The other person is a young woman who has just completed programmer training. I'm not sure just what she knows.

"Do you think you can get the system done in time if I put these other two people right on it and we just give Terri what we think she should have?"

Alan tried not to let his dismay show as he said, "Can I think about this until tomorrow? Maybe I can come up with some ideas for getting this finished as quickly as possible."

"Okay, but if we are to get these other two people, I need to talk to their manager right away. Oh, I did hear that the young woman really likes to use fourth generation languages."

Alan walked back to his desk thinking, "What luck! All I need is some guy with twenty years experience to come in and take over my project. I've got to come up with an alternative plan by tomorrow. Maybe the development center staff will have some ideas. I'll go talk to them."

Using a fourth generation language

The next morning Alan goes into his manager's office and says that he has an idea. "I think we can have the system done in two months and still do the prototyping. Here's how.

"Why don't you borrow the woman who has just finished programmer training? While I finish the detailed analysis and design the program structure, she can work with Terri and do the prototyping. Terri is knowledgeable enough to work with inexperienced people.

"Then we can go ahead and code the whole system in the fourth generation language. She can code while I develop the user's guide and the programmer's documentation. Everyone says that coding goes much faster with the new languages. Here's our chance to prove it. We can be done on time, have a well-designed and well-documented system, and do prototyping.

"Another benefit of using a fourth generation language is that we can leave the design of the report formats until last. As long as we know that we have the data when and where we need it, we could even produce *ad-hoc* reports for a while if we're pressed for time. At least we can be sure that we have all the processes in place to capture and store the data by the time Terri needs them. What do you think?"

Alan waits eagerly while his manager considers his suggestion. Finally, his manager responds, "Okay. I'm probably crazy to let you talk me into this, but we'll try it. I *would* like to prove that we can use these new techniques, even if I don't quite understand them yet. Are you sure that the whole system can be coded using a fourth generation language?"

"Yes, I talked to the people in the development center. They say that the only problem I may have is if I want some reports formatted in just a certain way. Because fourth generation languages are set processors rather than record processors, reports like the Needs Form and the course Catalog are difficult to produce in the format they are now. I'm sure we can print the information, but we may need to take another look at the exact format. I really don't see that as a major problem, though."

"All right. I'll see if I can get you the help you want in the next few days. Just be sure and let me know if you run into any problems that mean you might not be done in time. I don't want any surprises."

Alan thanks him and goes back to his desk thinking, "This had better work."

A couple days later, Alan's manager introduces Peggy to Alan. Much to his delight, she is proficient in FOCUS, one of the languages that International Telewidgets has installed. He gives her the partially completed Requirements Definition documentation to read and introduces her to Terri.

Alan finishes documenting the system inputs and outputs, describes the data stores (see Appendix B) and starts working on the detailed process model.

Questions

1. Is Alan smart to proceed with the detailed analysis and module design before the screen formats are defined? Why?
2. What tasks will Alan be doing in detailed analysis?
3. What did Alan forget to document? Why do you think he forgot it?

Exercises

Using the exercise you were given in Chapter 6, do the following:

1. List each data element in each system input and output.
2. List all the different transactions and assign an assumed volume to each. An example is a large, red, round transaction.

References

DeMarco, Tom. *Structured Analysis and System Specification.* New York: YOURDON Press, 1978.

Dickinson, Brian. *Developing Structured Systems.* New York: YOURDON Press, 1980.

8
The Detailed Process Model

> . . . even the best structured programming code will not help if the programmer has been told to solve the wrong problem, or, worse yet, has been given a correct description, but has not understood it. The results of requirements definition must be both complete and understandable.
>
> D. T. Ross and K. E. Schoman, Jr.,
> "Structured Analysis for Requirements Definition,"
> *Classics in Software Engineering.* New York: YOURDON Press, 1979, p. 365.

Alan is ready to explode the Level 1 data flow diagram, write the mini-specs, and add the data flow definitions to the data dictionary. Time has become a critical resource.

Exploding the data flow diagram

Alan figures that the only problem he will have with this phase of the project is knowing when he has exploded the data flow diagrams enough. He knows that he must partition the processes enough to be able to allocate them to subsystems. Subsystems are created when parts of the system are processed by different machines or different people.

He recalls that the general rule is to explode until he cannot further subdivide the process. "That's fine," he thinks, "but how do I know when I can no further subdivide the process?" He remembers reading a couple of hints: If the process includes both manual and mechanized parts, it is not exploded enough. If there is more than one input or more than one output from the process, it probably can be subdivided further. Since most people do not explode the processes far enough, he should keep going until there is one indivisible function or until the lower level diagram becomes more complex than the previous level.

He knows that some of the processes in the training system are manual and will remain manual, some are obvious candidates for mechanization, and some he may need to discuss with Terri. Once he defines a process that will stay completely manual, he does not need to explode that process further. However, he will want to write a process description for a manual process if clerical procedures may need to be written for it.

Design curriculum

The first major process to be exploded is "design curriculum" in Figure 8.1.

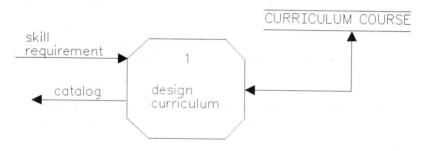

Figure 8.1.

According to the process notes, there are three major processes required to design the curriculum: the process of actually determining what changes need to be made to the curriculum as a result of Skill Requirements changing; the process of making the updates to the curriculum records; and the process of producing an updated course Catalog.

Alan draws the data flow diagram, Figure 8.2, that represents these processes.

THE DETAILED PROCESS MODEL   69

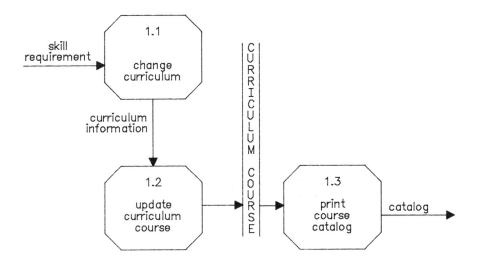

Figure 8.2.

Curriculum information contains *course description, course level, course name, course number, course prerequisite,* and *credit hours*. He adds the data flow *curriculum information* to the data dictionary, and writes process descriptions for these Level 2 processes.

Process descriptions commonly are called mini-specs and should be written in tight or structured English using data dictionary names for nouns and unambiguous verbs. Decision charts or decision trees also may be used if appropriate. Since valid values are part of the data dictionary, there is no need to repeat these in the process descriptions.

"Change curriculum" is a manual process for which manual procedures are written already, so Alan does not explode it further or write a process description for it.

"Update curriculum course" seems to be a logical candidate for mechanization, so he writes a process description for it.

Process Name: Update Curriculum Course
Process Number: 1.2
Process Notes:

Curriculum Course records may be added, changed, or deleted.

Can "update curriculum course" be subdivided further? Alan looks at his process notes. It appears that add, change, and delete are three separate functions. He explodes Process 1.2 in Figure 8.2 to show these three subprocesses. See Figure 8.3.

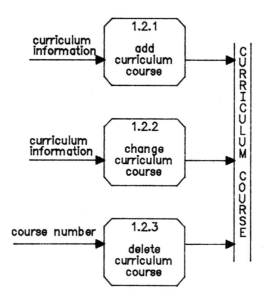

Figure 8.3.

Since it does not appear that any of these processes can be further subdivided, Alan writes a process description for each.

Process Name: Add Curriculum Course
Process Number: 1.2.1
Process Notes:

COURSE NAME, COURSE NUMBER, COURSE LEVEL, COURSE DESCRIPTION, and CREDIT HOURS all must be present and valid. COURSE PREREQUISITE if present must be a valid COURSE NUMBER.

Add valid CURRICULUM INFORMATION to the Curriculum Courses. Duplicate COURSE NUMBERS are invalid.

Process Name: Change Curriculum Course
Process Number: 1.2.2
Process Description:

If the COURSE NUMBER exists in the Curriculum Course records, any data item except COURSE NUMBER may be changed.

COURSE NAME, COURSE LEVEL, COURSE DESCRIPTION, and CREDIT HOURS all must be present and valid. COURSE PREREQUISITE, if present, must be a valid COURSE NUMBER.

Process Name: Delete Curriculum Course
Process Number: 1.2.3
Process Description:

Delete CURRICULUM COURSE INFORMATION for the specified COURSE NUMBER from the Curriculum Course records.

"Print course catalog" also seems likely to be mechanized. Because it appears to be only one function, Alan writes a process description for it.

Process Name: Print Course Catalog
Process Number: 1.3
Process Description:

CURRICULUM COURSE INFORMATION is printed for one COURSE NUMBER or for all COURSE NUMBERS.

Schedule training

"Design curriculum" has been exploded completely and the data flows entered into the data dictionary. Now Alan analyzes "schedule training" in Figure 8.4.

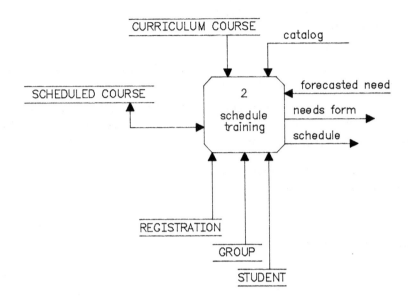

Figure 8.4.

According to the process notes, the processes required to schedule training are print a Needs form, maintain a record of Forecasted Needs, print accumulated needs, determine the class schedule, maintain a record of the Scheduled Classes, and print a class Schedule. See Figure 8.5.

Accumulated needs is composed of *course number, needed date,* and *number of needs. Scheduled course information* is composed of *course cost, course end date, course number, course start date, course time, instructor name,* and *room location.* Alan adds the data flows *accumulated needs* and *scheduled course information* to the data dictionary.

Alan analyzes the data flow diagram to see if there are any manual processes in "schedule training." All of these processes could be mechanized, but he assumes that he will not be mechanizing the process of actually determining the schedule, Process 2.4. All the other processes need to be exploded, if possible, and described.

"Print needs form" appears to be a single function with the following description:

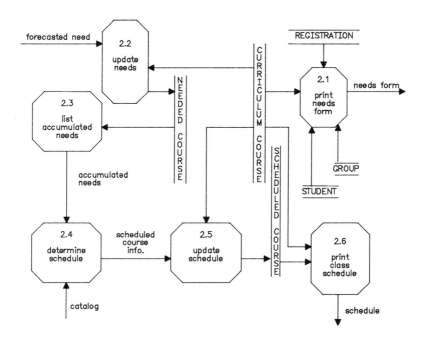

Figure 8.5.

Process Name: Print Needs Form
Process Number: 2.1
Process Description:

Print a form for each Student record. The form contains a list of COURSE NUMBER and COURSE NAME of all Curriculum Courses. COMPLETION DATE is printed for each completed course, and a space is provided for the student to indicate which Curriculum Courses he or she wishes to take and the preferred month.

Can update needs be exploded? Alan knows that whenever we are maintaining a data store, we have the possibility for adding, changing, or deleting. If the information being processed is very simple, we may not want to include the process of changing existing information. Instead we can simply delete and add. That is the way Alan will handle maintaining the Needed Course data store.

Process Name: Update Needs
Process Number: 2.2
Process Notes:

Needed Course records may be added or deleted.

Alan draws the Level 3 diagram, Figure 8.6, that shows these two processes.

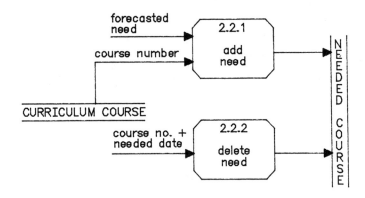

Figure 8.6.

Because neither of these two processes can be subdivided further, Alan writes the process descriptions.

Process Name: Add Need
Process Number: 2.2.1
Process Description:

COURSE NUMBER must match the COURSE NUMBER of a Curriculum Course. NEEDED DATE must be present and valid.

If there is no Needed Course record for the COURSE NUMBER and NEEDED DATE, add a Needed Course record.

If there is a Needed Course record for the COURSE NUMBER and NEEDED DATE, increase the NUMBER OF NEEDS by one.

Process Name: Delete Need
Process Number: 2.2.2
Process Description:

Delete Needed Course record for the specified COURSE NUMBER and NEEDED DATE.

Now Alan looks at "list accumulated needs." This process arranges the Forecasted Needs by *course number, needed date,* and *number of needs.* It appears, then, that the process just prints a report of accumulated needs and does not need to be exploded further. He writes the process description.

Process Name: List Accumulated Needs
Process Number: 2.3
Process Description:

List Needed Course records by NUMBER OF NEEDS (greatest first), COURSE NUMBER, and NEEDED DATE.

"Update schedule" again has the standard processes of adding, changing, and deleting.

Process Name: Update Schedule
Process Number: 2.5
Process Notes:

SCHEDULED COURSE INFORMATION may be added, changed, or deleted.

Alan draws the Level 3 Diagram, Figure 8.7, which shows these processes.

None of these processes can be exploded further, so Alan writes the process descriptions.

76 THE STRUCTURED SYSTEMS LIFE CYCLE: A CASE STUDY

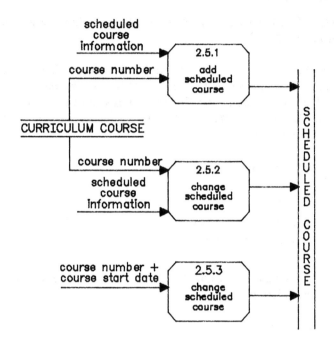

Figure 8.7.

Process Name: Add Scheduled Course
Process Number: 2.5.1
Process Description:

COURSE NUMBER must match the COURSE NUMBER of a Curriculum Course. COURSE START DATE, COURSE END DATE, COURSE TIME, INSTRUCTOR NAME, and ROOM LOCATION all must be present and valid.

Add valid SCHEDULED COURSE INFORMATION to the Scheduled Course records. Duplicate Scheduled Courses are not valid.

Process Name: Change Scheduled Course
Process Number: 2.5.2
Process Description:

If the COURSE NUMBER and COURSE START DATE exist in the Scheduled Course records, any other data item may be changed.

COURSE END DATE, COURSE TIME, INSTRUCTOR NAME, and ROOM LOCATION all must be present and valid.

Process Name: Delete Scheduled Course
Process Number: 2.5.3
Process Description:

Delete SCHEDULED COURSE INFORMATION for the specified COURSE NUMBER and COURSE START DATE.

Since Process 2.6 only prints the class Schedule, there is no need to explode it further.

Process Name: Print Class Schedule
Process Number: 2.6
Process Description:

Print SCHEDULED COURSE INFORMATION by SCHEDULED DATE and COURSE NUMBER if the COURSE START DATE is future.

Train students

"Train students" is the next process to be analyzed. See Figure 8.8.

78 THE STRUCTURED SYSTEMS LIFE CYCLE: A CASE STUDY

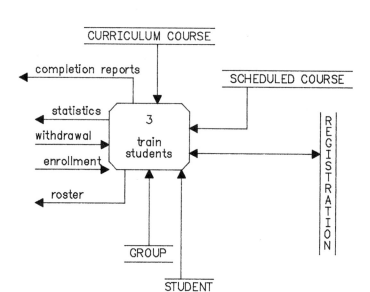

Figure 8.8.

Alan sees from the process notes that the processes required to present training are process Enrollments, process Withdrawals, teach, record Completions, report Completions, and report Statistics, shown in Figure 8.9.

THE DETAILED PROCESS MODEL 79

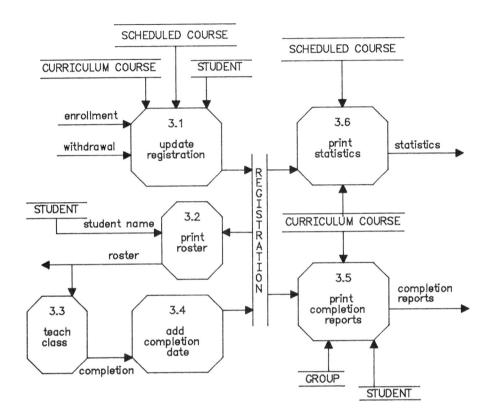

Figure 8.9.

Which of these processes will be mechanized and which will be manual? Teaching the class will stay manual, but the rest of them can be mechanized.

Process Name: Update Registrations
Process Number: 3.1
Process Notes:

ENROLLMENTS are added to the Registration records. Registration records are updated with WITHDRAWAL REASON. Registration records also may be deleted.

80 THE STRUCTURED SYSTEMS LIFE CYCLE: A CASE STUDY

Can this process be exploded further? Yes, it can because there are two different inputs which are processed differently, as in Figure 8.10.

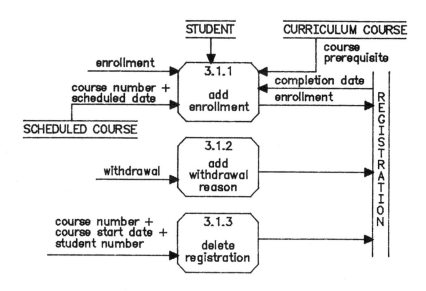

Figure 8.10.

Alan writes process descriptions (or notes) for these processes.

Process Name: Add Enrollment
Process Number: 3.1.1
Process Description:

STUDENT NUMBER must be in the Student records and COURSE NUMBER and COURSE START DATE must identify a Scheduled Course.

The Curriculum Course records are referenced to see if there is a COURSE PREREQUISITE for the course. If there is, then the Registration records are referenced to verify that the student has completed the prerequisite.

Add valid ENROLLMENT to the Registration records. Duplicate ENROLLMENTS are invalid.

Alan thinks that this process possibly could be exploded further into functions of editing the *student number, course number,* and *course start date*; editing the *course prerequisite*; and updating the Registration records. See Figure 8.11.

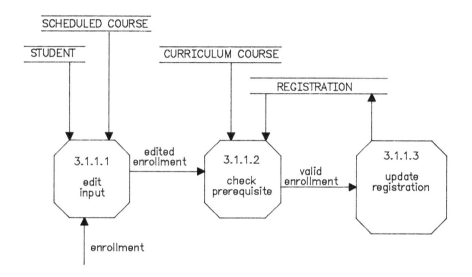

Figure 8.11.

He wonders which is the best. The argument against exploding is that the data going out of one of the subprocesses and into the next has not been changed in any way, only edited. On the other hand, if Process 3.1.1 contained more than a page of description, it might be better to explode it further. He decides to leave Process 3.1.1 unexploded.

He writes the process descriptions for Process 3.1.2 and Process 3.1.3.

Process Name: Add Withdrawal Reason
Process Number: 3.1.2
Process Description:

Add WITHDRAWAL REASON for the specified COURSE NUMBER, COURSE START DATE, and STUDENT NUMBER in the Registration records.

Process Name: Delete Registration
Process Number: 3.1.3
Process Description:

Delete Registration record for the specified COURSE NUMBER, COURSE START DATE, and STUDENT NUMBER.

Process 3.2, "print roster," only formats and prints, so it does not need to be exploded further.

Process Name: Print Roster
Process Number: 3.2
Process Description:

By COURSE NUMBER, print STUDENT NAME and COURSE NAME for each Registration record with a future COURSE START DATE.

Process 3.4, "add completion date," is a single function.

Process Name: Add Completion Date
Process Number: 3.4
Process Description:

Add COMPLETION DATE for the specified COURSE NUMBER, COURSE START DATE, and STUDENT NUMBER to the Registration records.

"Print completion report" only formats and prints.

Process Name: Print Completion Report
Process Number: 3.5
Process Description:

By GROUP NAME, print STUDENT NAME, COURSE NUMBER, COURSE NAME, COMPLETION DATE, and WITHDRAWAL REASON for each Registration record with a COURSE START DATE during the reporting period.

"Print statistics" is a single function.

Process Name: Print Statistics
Process Number: 3.6
Process Description:

By COURSE NUMBER and by INSTRUCTOR NAME, print COURSE NAME, COURSE DAYS, number of Registrations with non-blank COMPLETION DATE, and STUDENT DAYS for each Registration record with COURSE START DATE within the reporting period.

COURSE DAYS and STUDENT DAYS must be added to the data dictionary.

Maintain personnel information

The last major process to be analyzed is "maintain personnel information," diagrammed in Figures 8.12 through 8.15.

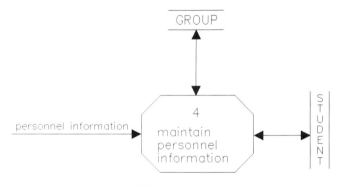

Figure 8.12.

This process maintains both Student and Group records, two separate processes that need to be drawn.

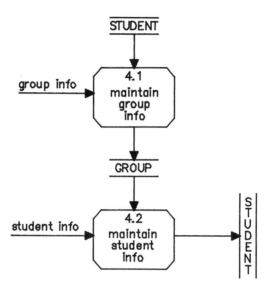

Figure 8.13.

Process Name: Maintain Group Information
Process Number: 4.1
Process Notes:

GROUP INFORMATION is added to, changed, or deleted from Group records.

This process can be exploded into three subprocesses.

THE DETAILED PROCESS MODEL 85

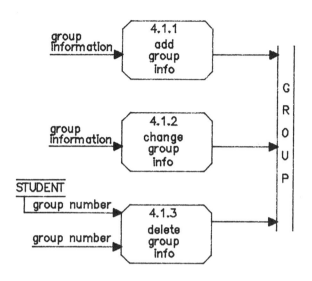

Figure 8.14.

Process Name: Add Group Information
Process Number: 4.1.1
Process Description:

GROUP MANAGER NAME and GROUP NUMBER must be present and valid.

Add valid GROUP INFORMATION to the Group records. Duplicate GROUP NUMBERS are not valid.

Process Name: Change Group Information
Process Number: 4.1.2
Process Description:

GROUP MANAGER NAME must be valid and may be changed for the specified GROUP NUMBER in the Group records.

86 THE STRUCTURED SYSTEMS LIFE CYCLE: A CASE STUDY

Process Name: Delete Group Information
Process Number: 4.1.3
Process Description:

Delete GROUP INFORMATION for the specified GROUP NUMBER in the Group records if there are no Student records with that GROUP NUMBER.

The process notes for "maintain student information" show that process also can be exploded further.

Process Name: Maintain Student Information
Process Number: 4.2
Process Description:

STUDENT INFORMATION is added to, changed, or deleted from Student records.

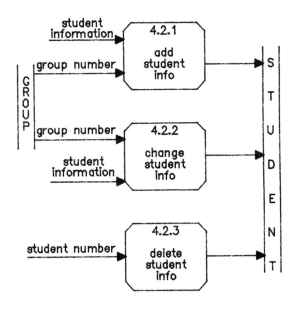

Figure 8.15.

Process Name: Add Student Information
Process Number: 4.2.1
Process Description:

STUDENT NAME, STUDENT NUMBER, and STUDENT LEVEL all must be present and valid. GROUP NUMBER must be a valid group.

Add valid STUDENT INFORMATION to the Student records. Duplicate STUDENT NUMBERS are not valid.

Process Name: Change Student Information
Process Number: 4.2.2
Process Description:

STUDENT NAME and STUDENT LEVEL must be valid and may be changed for the specified STUDENT NUMBER in the Student records.

Process Name: Delete Student Information
Process Number: 4.2.3
Process Description:

Delete the STUDENT INFORMATION from the Student records for the specified STUDENT NUMBER

All processes have been exploded fully and process descriptions (mini-specs) have been written for each lowest level process. The data dictionary includes all data items and data flows. Alan is preparing to have a walkthrough for Terri when his manager stops by.

"Well, Alan, how is the project going? Are we going to make the deadline?"

"It still looks like it," Alan replies. "The final version of the Requirements Definition document should be on your desk tomorrow. The prototyping worked beautifully. Terri had a lot a good ideas for the screen layouts that I wouldn't have thought of. And that software package we used to draw the data flow diagrams is great. The detailed analysis didn't take me nearly as long as I expected. I'm just getting ready to give Terri a walkthough of the process model to be sure I haven't left anything out."

"Good. Just keep me informed."

Holding a walkthrough

Alan goes to Terri's office and reviews the documentation with her once again. She makes notes as he is talking, and when he has concluded the walkthrough, he asks if she has noticed any problems.

"I did notice that you show a class schedule being printed, but not a schedule for each instructor and for each room. Was it too costly to do?"

"Well, uh, no. Actually, I forgot. It's no problem at all to add those. It's simply a different arrangement of information I'm already printing. I'll add those right away. Do you notice anything else I've left out?"

"Not really, but shouldn't you talk to the people in personnel? If they're going to be maintaining the personnel information, they might want to add something."

"Good suggestion. Thanks. I guess this is where my inexperience is showing. I'll talk to them."

"I can't think of anything else. Thanks to the prototyping, I can visualize how it's going to work, and I think it will be great. What do you have left to do?"

"I need to draw the structure charts, design the database, design the program modules and the code. Then we can code it. Based on what I saw with the prototyping, I don't think it will take long to test, so we should meet the deadline."

"Good. Let me know if you need me for anything more."

"I sure will. Thanks."

Making revisions

As Alan returns to his desk, he thinks about how easy it is to miss something and how important it is to review the documentation with the user periodically. He will revise the documentation to show that more than one schedule will be printed by Process 2.6 and then talk to the personnel people.

THE DETAILED PROCESS MODEL 89

Process Name: Print Schedules
Process Number: 2.6
Process Description:

Print SCHEDULED COURSE INFORMATION by SCHEDULED DATE and COURSE NUMBER, by INSTRUCTOR NAME, and by ROOM LOCATION, if the COURSE START DATE is future.

When Alan talks to the people in personnel, they are enthusiastic about the mechanized interface to the data processing training system. Their only request is to be able to get a listing of the personnel records whenever they request one.

Alan adds the process to produce the listing and defines the information requirements for the Personnel List. See Figure 8.16.

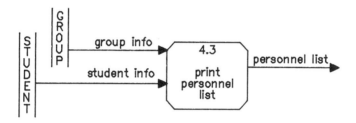

Figure 8.16.

Process Name: Print Personnel List
Process Number: 4.3
Process Description:

Print GROUP INFORMATION and STUDENT INFORMATION in sequence by GROUP NUMBER and STUDENT NUMBER.

Personnel List has an output volume of ten pages whenever requested. The major data elements are *group number*, *group manager name*, *student number*, *student name*, and *student level*. It is retained one cycle, and there are no special security requirements.

Process allocation

Since each subsystem in a system is designed independently, before Alan starts the design he needs to allocate processes to subsystems. He already has decided which functions will remain manual, but has not considered formally whether there is more than one mechanized subsystem.

Subsystem partitions are identified in several ways. Pieces of the system that will run on different computers will be allocated to different subsystems. An on-line part of the system probably would be in a different subsystem than the batch processes. Functions administered by different user organizations probably would be in different subsystems.

At first, Alan does not think that he has more than one subsystem. He is planning to code the whole system in a fourth generation language and run it on the mainframe. Terri wants at least part of it on-line, and the transaction volume is small enough that Alan has decided to make the entire system on-line.

Then, as he looks at the context diagram, he notices that there are two distinct groups of people who will be interacting with the system. One group is the students and their managers. The other group is the people in personnel who will be providing the personnel information. He decides to make Process 4, "maintain personnel information," a separate subsystem and will design and implement it first. While he designs the programs, Peggy can design the database.

Questions

1. What is a quick way to tell what level data flow diagram you are looking at?
2. Why is the detailed process model documentation not part of the Requirements Definition document?

Exercises

Using your answers to the exercises in Chapters 6 and 7, do the following exercises:

1. Explode the Context Diagram as far as possible, and evaluate the data flow diagrams according to the following rules.

   - All system inputs and their sources and all system outputs and their recipients should show on the context diagram.
   - As the diagrams are exploded, no data flows or data stores should be dropped, and the only ones that should be added are those which were totally inside the higher level process.
   - No data should flow to a process where it is not used.
   - Process names should consist of action verbs and objects.
   - No process should only rearrange data.

2. Write mini-specs and data dictionary definitions. (You may need to use your imagination.) See Appendix E for sample data element definition and sample data flow definition.

References

DeMarco, T. *Structured Analysis and System Specification.* New York: YOURDON Press, 1978.

Dickinson, Brian. *Developing Structured Systems.* New York: YOURDON Press, 1980.

Page-Jones, M. *The Practical Guide to Structured Systems Design.* New York: YOURDON Press, 1980.

Ward, Paul T. *Systems Development Without Pain.* New York: YOURDON Press, 1984.

9

The Design Model

> *... the design itself embodies its own understanding — to look upon it is to know what it can do, what it cannot do, and what we can do without destroying its essential nature .... It must encourage each hand to maintain not just its form, but its spirit. It will not only be maintained, but* its maintainability will be maintained.
> Gerald M. Weinberg, *Rethinking Systems Analysis and Design.*
> Boston: Little, Brown and Company, 1982, p. 102.

Alan is at his desk looking at the system development standards manual, trying to determine how to organize the rest of the development work, when his manager comes by to see how his work is progressing.

"Well, Alan, I noticed from your status report that you're a little behind schedule."

"That's true," Alan replies, "but I think we'll make it up next week. I just wish the Requirements Definition document would be approved. I hate to start into design until I know for sure what will be mechanized. I've been reviewing *The Practical Guide to Structured Systems Design* by Meiler Page-Jones while I wait."

"I don't think that there's any danger in proceeding. Terri tells me that she knows there won't be any problem because she's been keeping her boss informed, and he's been very enthusiastic about the project. It's just that he has been having so many meetings he hasn't had time to read the document.

"I've already signed it, and I'm sure if there are any changes, they'll be minor. In fact, I thought it was one of the best Requirements Definition documents I've ever read. My boss was really impressed that we had used prototyping. I also told him that we were going to code it in a fourth generation language. How will that affect the design effort?"

"I'm glad you were pleased with the document," Alan says. "I felt good about the way it turned out. Actually, the quality of the analysis documentation will have more impact on the design effort than will the use of a fourth generation language. The module hierarchy still needs to be designed, and coding in a fourth generation language doesn't have much impact on that.

"Most of the productivity gain from using a fourth generation language will be in coding and testing. For example, the code to describe the screen layouts will be generated from the screens that were designed for the prototype. Many of the mistakes programmers make, such as problems with first and last records, inconsistently defined data elements, and so on, simply don't happen in fourth generation languages. I think you'll be pleased at how quickly the system will be implemented."

"I'm certainly pleased so far, and so are our customers. Keep up the good work. By the way, Peggy will be able to help until the project is completed. You may want to let her do all of the coding while you do the testing and conversion."

"I'll do that. Let me know as soon as you hear anything about the Requirements Definition document approval."

One reason Alan is so confident that the design will not take very long is that he will be designing from fully exploded data flow diagrams. As a result, his first-cut structure charts should already be functionally cohesive. He still will check the completed structure charts for cohesion, but they should not need much revision.

He also knows that working from fully exploded data flow diagrams means that the only processes which will need to be added and defined during design are those processes which are concerned with the physical implementation of the system. These are the processes required for control, formatting, table handling, and file access. There should not be very many of these, particularly since he is going to code in a fourth generation language that generates much of the code for those processes. If he finds a need to add other types of processes, it means his data flow diagrams were either incomplete or not fully exploded and should be corrected.

Alan goes back to the systems development standards manual to make a list of the design activities. These activities include the following:

- transforming the expanded data flow diagram for each subsystem into a structure chart;
- adding control, formatting, and file access processes;
- designing logical and physical files or databases;
- packaging processes into modules and programs;
- designing the code.

Drawing the expanded data flow diagram

Next, Alan draws the expanded data flow diagram for the personnel subsystem, which is simply all the lowest level data flow diagrams connected together.

An expanded data flow diagram is a working document, so it does not need to be "pretty," but if it looks messy because of too many lines crossing, Alan will repeat a data store symbol where the data store is referenced, but not updated. If his expanded data flow diagram still looks like a bowl of spaghetti, he needs to work more on the process model. Difficulties in design are often attributable to sloppy analysis. Alan's expanded data flow diagram for the personnel subsystem looks like Figure 9.1.

Alan looks at the expanded data flow diagram and tries to determine whether it is a transform centered or transaction centered system. Transform centered systems are characterized by an input leg, a central transform, and an output leg. The input leg accepts system input as received from external sources, edits it, and puts it in a form ready for processing. The central transform consists of those processes which transform the edited input data into the output data. The output leg puts the processed data into system output format for the recipient.

Transaction centered systems are characterized by many inputs (or outputs), and possibly no central transform at all. The control process in a transaction center acts much like a traffic cop and routes each incoming transaction to its proper destination. Most systems are a combination of both of these design types. It is common to see transform centers within transaction centers, and vice versa.

Constructing the structure chart

Alan does not see any functions that seem to be transforming data. There are six different operations being performed on data stores and one obvious system output function. It appears this subsystem is primarily transaction centered. He draws a first-cut structure chart, being careful to include all the processes in the expanded data flow diagram, shown in Figure 9.1.

He adds one control process and calls it "route personnel transaction." See Figure 9.2.

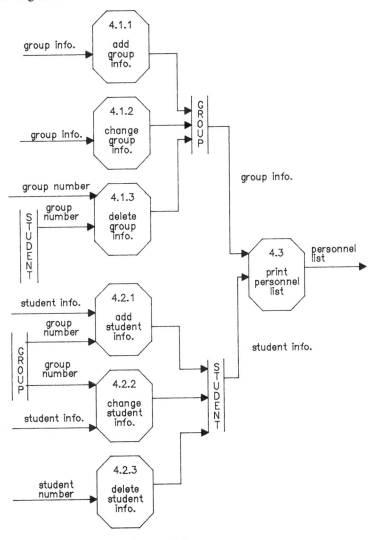

Figure 9.1.

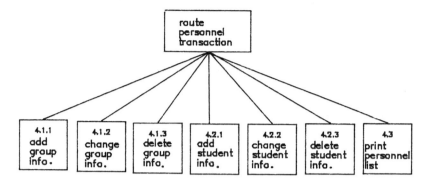

Figure 9.2.

To Alan, the structure chart looks strange because there is no process for accepting data that indicates which transaction to process. He looks at the screen layouts and sees that "selection code" is entered on the menu screen to show which type of transaction is to be processed. He adds a process to display the menu screen and accept the selection code. See Figure 9.3.

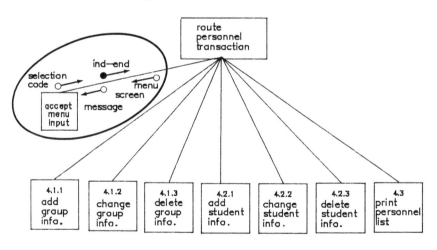

Figure 9.3.

He adds *selection code, ind-end,* and *message* to the data dictionary.

Selection Code:

The code which indicates which menu option has been selected.

Ind-End:

An indicator which is set when the last input screen has been received.

Message:

A response issued by the system indicating either successful execution or an error condition.

Next he adds the input and output processes for each transaction, shown in figures 9.4 through 9.9.

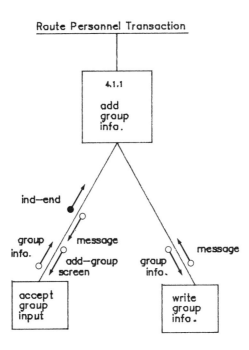

Figure 9.4.

98 THE STRUCTURED SYSTEMS LIFE CYCLE: A CASE STUDY

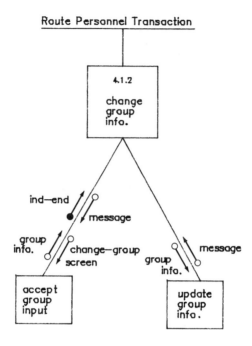

Figure 9.5.

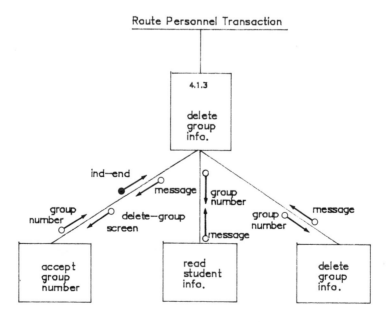

Figure 9.6.

THE DESIGN MODEL 99

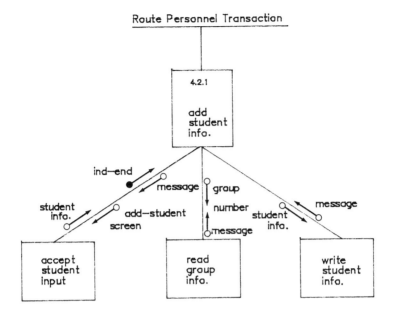

Figure 9.7.

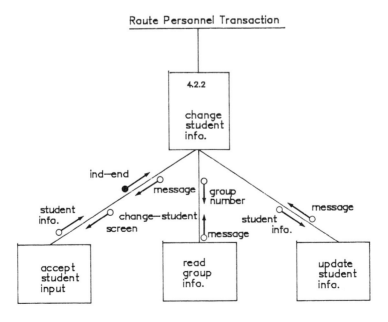

Figure 9.8.

100 THE STRUCTURED SYSTEMS LIFE CYCLE: A CASE STUDY

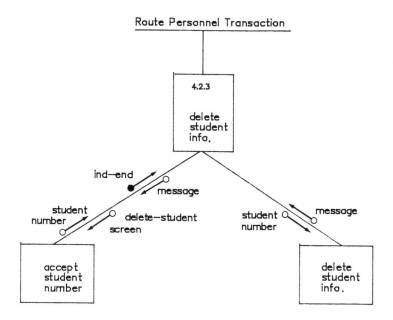

Figure 9.9.

Alan starts to add a formatting process to Process 4.3, "print personnel list," when he remembers that it will be coded in FOCUS. Thus, no file access modules will be coded, and the formatting will consist of one statement. He decides to include the formatting logic in Process 4.3, shown in Figure 9.10.

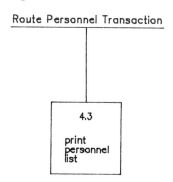

Figure 9.10.

Alan's structure chart for the personnel subsystem looks like Figure 9.11.

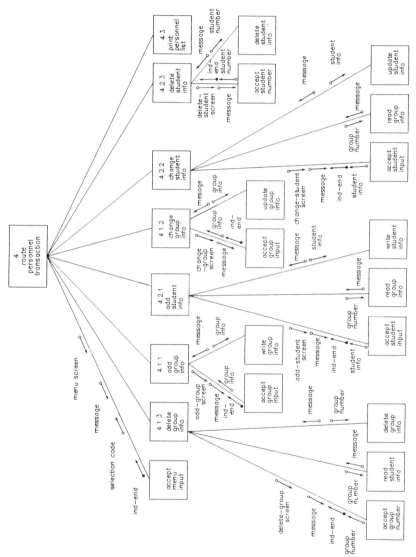

Figure 9.11.

Evaluating the structure chart

As Alan reviews his structure chart, he evaluates it according to four criteria:

1. High Cohesion. The highest type of cohesion is functional cohesion, in which each process is only one indivisible function. Structure charts done from expanded data flow diagrams are by definition functionally cohesive, but should be re-evaluated.
2. Low Coupling. Very little data connects the processes.
3. No split decisions. The process that makes a decision also has the data required to make the decision.
4. Fully factored model. Transaction centers will look somewhat flat if there are many transactions, but only that portion of the model should look flat. If a process has more than nine subordinate processes, another control level probably should be added.

Alan thinks that his design model looks quite good. It is easy to understand; it matches the real-world system; there is not a lot of data being passed; the control function is making decisions; and no process has too many subordinates.

He wonders what the structure would look like if he had put all the input and output processes under control of the main process instead of under the individual transactions, as in Figure 9.12.

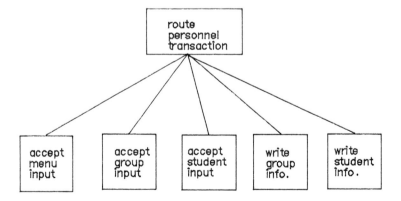

Figure 9.12.

The structure would be a complete "pancake." The main function would have to make every decision as well as the validation of data for each transaction. If he added an edit function for each transaction, as in Figure 9.13, the main function would have to make an additional decision for each transaction, and the data would have to be passed more than in his original design. He decides that his first choice, 9.11, was the best.

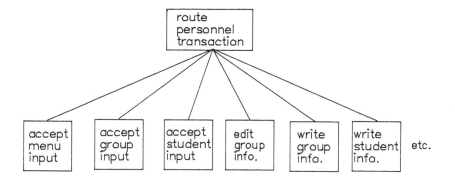

Figure 9.13.

Correcting an omission

Alan is filing his documentation when his phone rings. It is Terri.

"Hi, Alan. My boss finally had time to read the Requirements Definition document. He says it's the best one he has ever read. Just one small omission. He noticed where the amount for chargeback was entered on the screen and printed on the report, but couldn't find it on the information model. I'm surprised he caught that, but he said the information model was the easiest model for him to understand."

"Oh, I should have caught that. When my manager told me about adding that requirement, I was so concerned with the change in due date that I forgot to add the new requirement. I suppose that this will delay the approval some more."

"No, he just wrote a note when he signed the document, so everything is approved."

"Good. I never realized approvals could take so long. Did you say that you wanted to charge according to the curriculum course or the scheduled course?"

"According to the scheduled course," Terri answers.

"Okay, I'll correct that. I have the design done for the personnel subsystem. It should be coded next week. Looks like we'll make the schedule with time to spare unless something unexpected comes up."

"Great. Talk to you later, Alan."

"Forgetting to add the course cost to the information model was really dumb," Alan thinks. "I guess that's why we have people review the documents, but it sure is embarrassing. Well, I had better get the personnel subsystem packaged so it can be coded."

Questions

1. At what point in the development life cycle does it matter who or what will perform a particular function?

2. How does the use of a fourth generation language affect design?

3. How is design affected by the analysis phase output?

4. Why would you want more than one subsystem in a system?

5. What type of processes are added during design?

6. Could Alan have coded directly from his data flow diagrams rather than doing design? Why?

Exercises

Using the results of the exercises in Chapter 8, do the following exercises.

1. Indicate the input and output streams and the central transform on the expanded data flow diagram.
2. Draw the structure chart.
3. Which processes would not need to be added if the system were going to be coded in a fourth generation language?

References

Page-Jones, M. *The Practical Guide to Structured Systems Design.* New York: YOURDON Press, 1980.

Weinberg, Gerald M. *Rethinking Systems Analysis and Design.* Boston: Little, Brown and Company, 1982.

10
The Packaged Model

> *I suggest that we confine ourselves to the design and implementation of intellectually manageable programs.*
> E.W. Dijkstra, "The Humble Programmer," *Classics in Software Engineering.*
> New York: YOURDON Press, 1979, p. 113.

Alan is sitting at his desk talking to his manager.

"I'm glad the project is coming along so well, Alan. Our use of a fourth generation language is really getting attention. What's left to do?"

"Well," Alan replies, "the databases are designed, but I still need to package the personnel subsystem processes into modules and programs and design the code. While that subsystem is being coded, I'll design the other subsystem. After it's coded, we can system test, convert the history file, and install the system. I don't see any problem in making our due date."

"I must admit this project has changed my opinion regarding structured techniques. We've had less rework and faster development on this project than any other I've been involved with. And I know it isn't due to your great experience! The real test will be when the system needs maintenance, though. By the way, what do you mean by package? Isn't each process a module?"

"No," Alan explains, "some of the processes are quite small. Having too many small modules can be difficult to keep track of in a maintenance environment. Sometimes too many calls can cause unacceptable response times as well. Packaging is sort of a balancing act between excessively small modules and excessively large modules. Of course, the type of processor to be used makes a difference. A small module for a mainframe may be excessively large for a PC."

Grouping processes into modules

"Is there any special method you use to decide how to package the processes?" his manager asks.

"We use procedural analysis." Seeing his manager's puzzled look, Alan continues. "All that means is that we make the packaging decisions based on an analysis of the procedural characteristics of the processes.

"We note what percentage of the time each process will be executed so that we can put seldomly executed processes in a separate module. We note any process which is invoked from more than one other process so that we can keep it in a separate module. We note where the loops are so that we keep parts of a loop in the same module where possible. We note which processes are executed together so that we can package them together, unless the module becomes unacceptably large. We note the approximate amount of logic to be coded for each process so that we can keep each module a reasonable size considering the environment under which it will run. We also need to keep high cohesion and low coupling between the modules."

Grouping modules into programs

"How do you know which modules to include in the same program?" his manager asks.

"There are a couple of considerations when deciding on programs," Alan answers. "As a general rule, programs are separated by permanent files, but there are other considerations. If certain processes can be executed the same time as other processes, putting them in different programs can cut down on the elapsed time of a run. Sometimes processes are put into separate programs because of backup and recovery considerations or security considerations. The personnel subsystem is fairly small and only uses two related data stores, so it will probably be one program. The other subsystem may consist of several programs, however."

"Well, it sounds like you know what you are doing. I'll talk to you later."

Using procedural analysis

Alan takes a copy of the personnel subsystem structure chart and makes notes on it indicating size, frequency, and loops in Figure 10.1.

Alan first looks to see if there are any processes which are executed very infrequently. These are typically initialization or error routines. He does not see any of those.

Next, Alan checks to see if any processes are initiated from more than one other process. These are normally processes such as file access routines and table handlers. He does find several common file access processes: "read student information" and "read group information." If the system were going to be coded in a language such as COBOL, he would want to make these file access processes separate modules and call them. Since this system is going to be coded in a fourth generation language, there is no need for that because the programmer does not code file accesses.

Now Alan looks for processes that always execute together. Only if the resultant module would be too large for the type of physical implementation chosen would he separate functions that always execute together. Alan notices that the file access processes for each transaction always execute together, so he will leave the file access processes with their parent processes. See Figure 10.2.

Because the control process and the process of accepting input from the menu screen always execute together, Alan packages them together. None of the other processes execute together because once it is determined from the menu which transaction is to be processed, control does not return until all transactions of that type are completed.

Figure 10.3 shows how Alan packaged the processes. Depending on the physical implementation of a particular system, there could be many different correct ways of packaging the processes. Procedural analysis is a way to consider the issues in a logical manner.

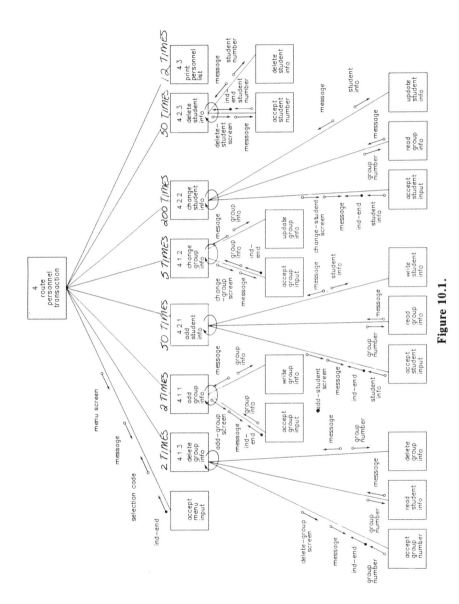

Figure 10.1.

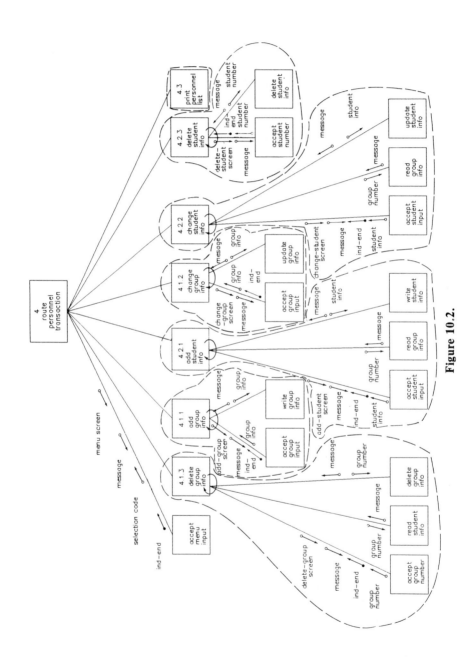

Figure 10.2.

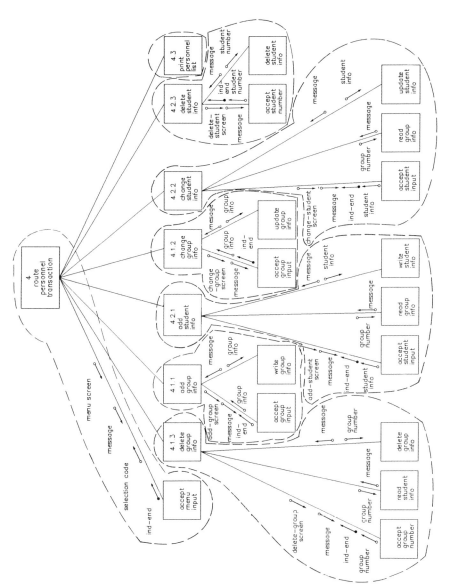

Figure 10.3.

112 THE STRUCTURED SYSTEMS LIFE CYCLE: A CASE STUDY

Tuning for efficiency

If there are any efficiency problems after the system is operational, Alan will analyze where the problems seem to be occurring and concentrate on tuning only those areas since any further packaging could adversely impact the maintainability of the system.

Designing the code

Alan is now ready for the last step in design: code design. For this procedure he uses Nassi-Shneiderman (N/S) charts. He could have used other methods such as pseudocode, but the system development standards manual recommends the use of N/S charts. Alan designs the module which contains the control process and the main menu process in Figure 10.4. See Appendix D for the rest of the N/S charts.

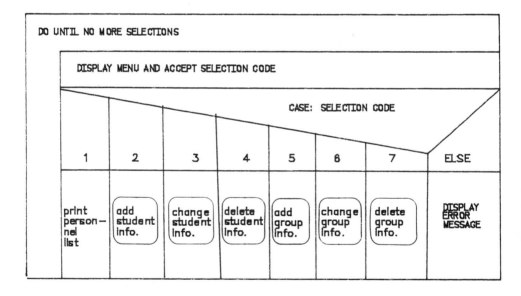

Figure 10.4.

Alan has completed design of the personnel subsystem. He files the structure chart, procedural analysis notes, and N/S charts in the design document and starts designing the rest of the system.

Questions

1. What is packaging?
2. List the steps in procedural analysis.
3. What criteria do you use to group modules into programs?
4. At what point in the development life cycle do you consider the efficiency of the code?

Exercises

1. Use procedural analysis on the structure chart that you produced in the exercises in Chapter 9. Group the processes into modules and explain your rationale. Assume that the system is at least three modules and that 20 of the 120 transactions will reject because of error.
2. Draw an N/S chart for each module. Refer to the process descriptions in Chapter 8, Exercise 2.

References

Booth, Grayce M. *The Design of Complex Information Systems.* New York: McGraw-Hill, 1983.

Page-Jones, M. *The Practical Guide to Structured Systems Design.* New York: YOURDON Press, 1980.

11

The Maintenance Request

A rigorous design should survive its implementation, not be swamped by it, and provide a framework for the intellectual control of changes to the implementation as requirements change.

Harlan D. Mills, *Software Productivity*.
Boston: Little, Brown and Co., 1983, p. 239.

Alan is sitting at his desk trying to find where to make a maintenance change in a program that was not designed using structured techniques when his manager stops by.

"Having trouble with that change request, Alan?" his manager asks.

"I sure am. Don't you have any more development projects for me to work on?"

"No, not right now. Don't worry though. Just as soon as I get one, you'll get it. Everyone was very impressed with your work on the training system. Speaking of which, Terri has a change request. Give her a call when you get a minute."

Alan makes the change to the program he has been working on and calls Terri.

"Alan, I'm glad you called. Our new training system is exactly what I asked for, but there's just one thing we would like to add — automatic scheduling."

"How soon do you need it?"

"No rush. We've gotten by without it for a long time, and accumulating and listing the needs has helped me do the scheduling faster. But it really would save me some time," Terri says.

"Is that why you kept mentioning the instructor skills and room facilities when I was first determining the requirements for the system?" Alan asks.

"Yes. I was trying to decide whether or not to ask you to include automatic scheduling then. I decided I would wait and see how the rest of the system turned out before I asked for that. But since you did such a good job, I figured you could handle the automatic scheduling process as well. Will it be too much of a problem to add now?"

"It shouldn't be. If you can tell me exactly what thought process you go through to determine the schedule, adding it to the system should be no problem."

Analyzing the change

Alan remembers that he included all the functions when he analyzed the system, even those that were not going to be mechanized. He goes back and looks at the data flow diagrams and sees that Process 2.4, "determine schedule," already is partially defined, as shown in Figure 11.1.

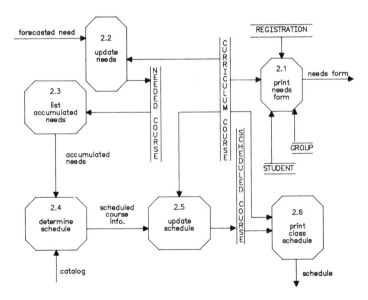

Figure 11.1.

116   THE STRUCTURED SYSTEM LIFE CYCLE: A CASE STUDY

Alan has never made a maintenance change to a system that has structured documentation, and he's not quite sure where to start. He looks to see if the systems development standards manual covers maintenance, but for maintenance it only mentions following the development steps, starting at the point where something changes. "Okay," he thinks, "I'll start reading through my documentation until I find where I need to make a change."

Revising the information model

There is no change in scope because scheduling was already inside the system. He needs to make some slight changes to the information model. He makes INSTRUCTOR and ROOM objects. He adds *instructor availability, instructor skill level, room size, room availability,* and *room facilities* to the data dictionary and attributes them to their respective objects. *Instructor name* and *room location* are removed from the object SCHEDULED COURSE and attributed to INSTRUCTOR and ROOM. The revised information model looks like Figure 11.2.

Alan writes the object definitions for INSTRUCTOR and ROOM and the relationship definitions for "instructs" and "is taught by."

Figure 11.2.

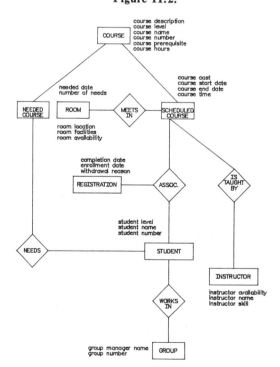

Revising the process model

He now has two additional data stores, Instructor Information and Room Information, included in Process 2.4. Since the scheduling now will be mechanized, *curriculum course information* will need to come from the data store rather than the course Catalog. See Figure 11.3.

Alan follows the rest of the development steps just as he did the first time. He describes the two new data stores and revises the description of Scheduled Course. He explodes the data flow diagram for "determine schedule," writes the process descriptions, documents the two new system inputs, makes a revised expanded data flow diagram for that part of the system, revises the structure charts, perhaps repackages into modules, draws N/S charts, codes, tests, and writes procedures.

"Now this kind of maintenance isn't bad at all," Alan decides. "In fact, it's fun. It's like adding a vertical slice to the development documentation. This documentation is in just as good shape as when I finished development." He calls Terri to tell her that she can use the automatic scheduling process now.

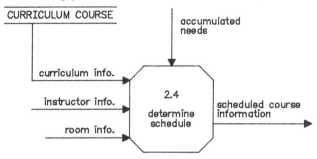

Figure 11.3.

"Finished already, Alan? I thought that it would take months. You must have worked a lot of overtime to get that done so fast."

"No. We've just begun to reap some of the benefits from a well-designed, well-documented system. I just hope my manager doesn't give it to someone else to maintain who won't follow the design or maintain the documentation. Just a few poorly made changes could make this system just as hard to maintain as the one I was just working on."

"Well, maybe he'll turn it over to someone who has learned structured techniques in college like you did. And we're offering some pretty good in-house courses now."

"Maybe. Anyway, I have certainly enjoyed working on this system. Talk to you later."

118 THE STRUCTURED SYSTEM LIFE CYCLE: A CASE STUDY

As Alan prepares to work on his next assignment, he thinks about what he has accomplished with the training system project.

- He produced the system on time.
- The customer is very happy with it.
- The system is easy to maintain.
- His manager is pleased with his work.
- His manager now has a positive attitude towards structured techniques. He has seen how using structured techniques force developers to follow the most direct path to a good implementation.
- He has learned a lot about how the structured techniques are applied within a development life cycle.

All in all, it was a good first development project, and he is ready to tackle something bigger.

Questions

1. What is the general rule about the steps to be followed for maintenance changes?
2. How does structured development impact the maintenance effort?

Exercise

1. In the problem you have used in these exercises, explain what would need to be changed under the following circumstances:

    a. If green transactions were now allowed?

    b. If the calculation factors in the table were changed?

    c. If large transactions were to be written on a separate file from small transactions?

References

Page-Jones, M. *The Practical Guide to Structured Systems Design.* New York: YOURDON Press, 1980.

Carma McClure, Carma. *Software Maintenance.* Englewood Cliffs: Prentice-Hall, Inc., 1983.

Appendix A

Manual System Sample Forms and Documents

Table of Contents

Completion Form	121
Completion Report	121
Course Description	122
Course Roster	122
Course Schedule	122
Enrollment Request	123
Forecasted Needs Report	123
Needs Form	124
Room Schedule	124
Training Statistics	125

Appendix A

COMPLETIONS

```
                    COMPLETIONS
    COURSE #    C213       COURSE NAME   ADVANCED COBOL
    COURSE DATE 1-84       INSTRUCTOR    SMITH


    STUDENT #      STUDENT NAME    COMPLETED      NO-SHOW

    111-11-1111    BROWN B         1-5-84
    222-22-2222    COOK C                         1-5-84
    333-33-3333    DAVIS D         1-5-84
```

DP TRAINING REPORT FOR JANUARY 1984

```
                 DP TRAINING REPORT FOR JANUARY 1984

    GROUP #: 1     GROUP MANAGER: ADAMS

    STUDENT NAME    COURSE #/NAME         COMPLETION DATE   NO-SHOW DATE

    BROWN B         C213 ADVANCED COBOL   1-5-84
                    C300 DATA GATHERING                     1-5-84

    JONES J         C300 DATA GATHERING   1-11-84


    GROUP #: 2     GROUP MANAGER: WEEKS

    STUDENT NAME    COURSE #/NAME         COMPLETION DATE   NO-SHOW DATE

    COOK C          C213 ADVANCED COBOL                     1-5-84
    DAVIS D         C213 ADVANCED COBOL   1-5-84
    JONES J         C300 DATA GATHERING   1-11-84
    SMITH S         C300 DATA GATHERING   1-11-84
```

COURSE DESCRIPTION

```
          COURSE DESCRIPTION

  COURSE NAME:              COURSE #:
  COURSE LEVEL:             CREDIT HOURS:
  COURSE DESCRIPTION:




  COURSE PREREQUISITE:
```

JANUARY COURSE ROSTER

```
          JANUARY COURSE ROSTER

  C213  Advanced Cobol

        Brown B
        Cook C
        Davis D

  C300  Data Gathering

        Brown B
        Jones J
        Smith S
```

COURSE SCHEDULE

	COURSE SCHEDULE					
	JANUARY 1984					
COURSE #	COURSE NAME	HOURS	DATE	TIME	RM #	INSTR
C213	ADVANCED COBOL	24	1/3–1/5	8:30–4:00	450	SMITH
C300	DATA GATHERING	16	1/10–1/11	8:30–4:00	448	JONES
C120	SPF	40	2/6–2/10	8:30–4:00	450	DAVIS

ENROLLMENT REQUEST

```
                    ENROLLMENT REQUEST
            COURSE #      COURSE NAME      COURSE  DATE
  ADD       _____     _____      _____

            _____     _____      _____
  DELETE    _____     _____      _____


  STUDENT NUMBER   _____
  STUDENT NAME     _____
  DATE             _____
```

FORECASTED NEEDS REPORT

COURSE #	COURSE NAME	QTR	# NEEDS
	FORECASTED NEEDS REPORT		
	1984		
C213	Advanced Cobol	1	3
C300	Data Gathering	1	2
C300	Data Gathering	2	1
etc.			

NEEDS FORECAST 1984

TRAINING NEEDS FORECAST 1984   JONES J   LEVEL: 2
MANAGER NAME: WEEKS

COURSE #	COURSE NAME	COURSE LEVEL	COMPLETED	DESIRED QTR
C100	xxxxxxxxxx	1	1-15-83	
C110	xxxxxxxxxx	1	2-20-83	
C120	xxxxxxxxxx	1	4-10-83	
C200	xxxxxxxxxx	2	9-20-83	
C213	ADVANCED COBOL	2		1/84
C240	xxxxxxxxxx	2		3/84
C300	DATA GATHERING	3		1/84
C320	xxxxxxxxxx	3		4/84

ROOM SCHEDULE

Room 450
January 1984

MON	TUE	WED	THUR	FRI
2	3 C213	4 C213	5 C213	6
9	10	11	12	13
16	17	18	19	20
23	24	25	26	27
30	31			

TRAINING STATISTICS

```
                TRAINING STATISTICS

    Course #     # Students    Student Days    Instructor

    C213             2             6             Smith
    C300             2             4             Jones
                    ───           ───
                     4            10
```

Appendix B

Requirements Definition

Table of Contents

System Overview	128
Problem Definition	128
Business Objectives	128
System Objectives	128
Data Conversion Considerations	128
User Needs	129
Context Diagram	130
Event List	131
System Input Requirements	132
Enrollment	132
Forecasted Need	132
Personnel Information	132
Skill Requirement	133
Withdrawal	133
System Output Requirements	134
Completion Reports	134
Catalog	134
Needs Form	135
Personnel List	135
Roster	136
Schedules	136
Statistics	137

Information Model	138
Object-Relationship Diagram	138
Object Definitions	139
Course	139
Group	139
Needed Course	139
Registration	139
Scheduled Course	140
Student	140
Relationship Definitions	140
Associates	140
Needs	141
Works In	141
Data Store Descriptions	142
Curriculum Courses	142
Group Information	142
Scheduled Courses	143
Student Information	143
Registrations	144
Process Model	145
Level 1 Data Flow Diagram	145
Process Notes	146
Design Curriculum	146
Schedule Training	146
Train Students	146
Maintain Personnel Information	147

System Overview

Problem definition

The data processing training area keeps records about requested training, curriculum courses, scheduled courses, enrollments, completions, training resources, and students. It produces reports that provide information about enrollments, forecasted needs, completions, scheduled courses, and resource usage. In addition, it produces a course catalog.

The records are being kept manually on the word processor. As a result, they are not consistent, updating them is very time consuming, and the paper work is becoming unmanageable.

Business objectives

1. Eliminate the need for two clerical positions by providing complete and consistent training records and by minimizing the time and effort required to update them.
2. Save 25 percent of the registrar's time by providing quick and flexible reporting.

System objectives

1. Provide consistent and complete training records by updating each record with one transaction.
2. Minimize the time and effort required to keep training records by providing on-line updating and viewing capabilities.
3. Provide flexibility in reporting by providing *ad-hoc* reporting capabilities.

Data conversion considerations

Historical training completion records will need to be established on the mechanized system.

User needs

The major users of the data processing training system are the students and their managers. The personnel department is the source of personnel information. Training statistics are required by the data processing training manager. The following were identified as user needs:

1. Update all completion records with one transaction.
2. Keep a record of incompletes.
3. Record withdrawals by date and reason.
4. Make one entry to record an enrollment.
5. Report completions and withdrawals by time period by individual, group, course, and instructor.
6. Produce course rosters by month.
7. Produce a forecasted needs turnaround document.
8. Mechanically generate a course schedule.
9. Provide the capability for students to enter their own enrollments on-line.
10. Validate prerequisites at enrollment time.
11. Mechanically produce the course catalog.
12. Generate training statistics regarding student days and room usage.

Context Diagram

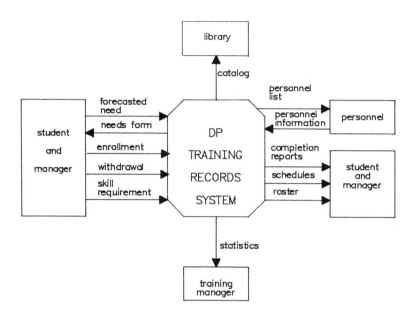

Event List

EVENT	SYSTEM INPUT	EXPECTED RESPONSE
Job skill requirements change.	Skill Requirements	Curriculum is designed.
Student foresees training need.	Forecasted Need	Curriculum courses are scheduled.
Student decides to take a course.	Enrollment	Student is trained.
Student decides not to take a course.	Withdrawal	Student enrollment is cancelled.
Personnel change occurs.	Personnel Information	Student and/or group information is changed.

System Input Requirements

Enrollment

Source: Student and Manager

Volume: 50 to 100 per month

Frequency: As required

Major Data Elements:

> Course number
> Course start date
> Enrollment date
> Student number

Forecasted Need

Source: Student and Manager

Volume: 200 to 500

Frequency: Annually

Major Data Elements:

> Course number
> Needed date

Personnel Information

Source: Personnel

Volume: 300 per month

Frequency: Monthly

Major Data Elements:

> Group number
> Group manager name
> Student level
> Student name
> Student number

Skill Requirement

Source: Manager

Volume: 2 to 5 per year

Frequency: Monthly

Major Data Elements:
>N/A
>
>(Skill requirements vary by student level.)

Withdrawal

Source: Student and Manager

Volume: 10 to 20 per month

Frequency: As required

Major Data Elements:
>Course number
>Course start date
>Student number
>Withdrawal reason

System Output Requirements

Completion Reports

User: Manager and Student

Purpose: Report training activity year to date

Security: None

Volume: 8 pages

Frequency: Monthly

Retention: None

Major Data Elements:

>Completion date
>Course name
>Course number
>Course start date
>Group manager name
>Student name
>Withdrawal reason

Catalog

User: Library

Purpose: Describe the data processing curriculum

Security: None

Volume: 1 to 200 pages

Frequency: As required

Retention: One cycle

Major Data Elements:

>Course description
>Course level
>Course name
>Course number
>Course prerequisite
>Credit hours

Needs Form

User: Student

Purpose: To provide a turn-around document on which to indicate forecasted training needs

Security: None

Volume: 300 pages

Frequency: Semi-annually

Retention: One cycle

Major Data Elements:

    Completion date
    Course name
    Course number
    Group manager name
    Student name

Personnel List

User: Personnel

Purpose: To provide a listing of the current group information and student information.

Security: None

Volume: 10 pages

Frequency: On request

Retention: One cycle

Major Data Elements:

    Group number
    Group manager name
    Student level
    Student name
    Student number

Roster

User: Manager and Student

Purpose: List the students enrolled in each scheduled course.

Security: None

Volume: 2 to 4 pages

Frequency: Monthly

Retention: 3 months

Major Data Elements:

    Course name
    Course number
    Course start date
    Student name

Schedules

User: Manager and Student

Purpose: List the data processing courses scheduled for the year

Security: None

Volume: 2 to 10 pages

Frequency: Monthly

Retention: 3 months

Major Data Elements:

    Course cost
    Course end date
    Course name
    Course number
    Course start date
    Course time
    Instructor name
    Room location

Statistics

User: Training manager

Purpose: Report training resource usage

Security: None

Volume: 2 pages

Frequency: Quarterly

Retention: 1 cycle

Major Data Elements:
        Course name
        Course number
        Course start date
        Number of students
        Number of student days
        Instructor name
        Room location

138 THE STRUCTURED SYSTEM LIFE CYCLE: A CASE STUDY

Information Model

Object-Relationship Diagram

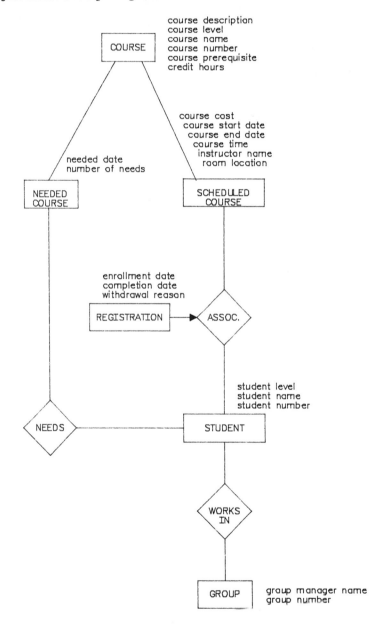

Figure B.2.

Object Definitions

Object: Course

A course offered by the data processing training group.

Attributes:

Course description
Course level
Course number
Course name
Course prerequisite
Credit hours

Object: Group

A work group in which a student works.

Attributes:

Group number
Group manager name

Object: Needed Course

A course for which a student has expressed a need.

Attributes:

Needed date
Number of needs

Object: Registration

A record of a student's registration in a scheduled course.

Attributes:

Completion date
Enrollment date
Withdrawal reason

Object: Scheduled Course

A course which has been assigned dates, room, and instructor.

Attributes:

Course cost
Course end date
Course start date
Course time
Instructor name
Room location

Object: Student

Someone who needs a course or enrolls in a scheduled course.

Attributes:

Student level
Student name
Student number

Relationship Definitions

Relationship: Associates

For each registration there is a scheduled course and a student such that these objects are associated when a student enrolls in a scheduled course.

Reverse Relationship:

1. For each student there is none, one, or many registrations.
2. For each scheduled course there is none, one, or many registrations.

Relationship: Needs

For each needed course there is a student such that each needed course is needed by one or many students.

Reverse Relationship:

Each student has none, one, or many needs.

Relationship: Works In

For each student there is a group such that each student works in one group.

Reverse Relationship:

Each group may contain one or many students.

Data Store Descriptions

Curriculum Courses

Security: Read only except by data processing training group

Currency: As of month end

Volume: 300 records

Retention: Permanent

Major Data Elements:

        Course description
        Course level
        Course name
        Course number
        Course prerequisite
        Credit hours

Group Information

Security: Read only except by data processing training group

Currency: As of month-end

Volume: 30

Retention: Permanent

Major Data Elements:

        Group number
        Group manager name

Scheduled Courses

Security: Read only except by data processing training group

Currency: As of month-end

Volume: 50 to 100 per year

Retention: Permanent

Major Data Elements:

>Course cost
>Course end date
>Course number
>Course start date
>Course time
>Instructor name
>Room location

Student Information

Security: Read only except data processing training group

Currency: As of month-end

Volume: 300

Retention: Permanent

Major Data Elements:

>Student level
>Student name
>Student number

Registrations

Security: None

Currency: As of month-end

Volume: 20 to 100 per month

Retention: Permanent

Major Data Elements:
>Course number
>Course start date
>Completion date
>Enrollment date
>Student number
>Withdrawal reason

APPENDIX B — REQUIREMENTS DEFINITION 145

Process Model

Level 1 data flow diagram

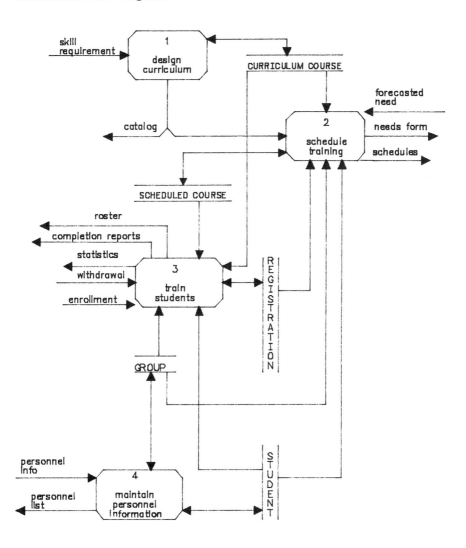

Process Notes

Process Name: Design Curriculum

Process Number: 1

Process Notes:

When Skill Requirements change, the curriculum is analyzed to determine if courses should be added or dropped from the curriculum or if courses should be modified. After these determinations are made, Curriculum Course records are updated and revisions to the course Catalog are printed.

Process Name: Schedule Training

Process Number: 2

Process Notes:

A Needs form is printed for each student. Each form contains a list of all the Curriculum Courses, showing which ones have been completed and giving a place to show when the student foresees a need to take any uncompleted courses.

When Forecasted Needs are received from the students, the total number of students expressing a need for each course is printed by NEEDED DATE.

The courses are scheduled according to the greatest need and the available room and instructor resources. The course Catalog is used to determine the course length and instructor requirements for the course. Scheduled Course records are updated and a revised class Schedule is printed.

Process Name: Train Student

Process Number: 3

Process Notes: Students enroll in and withdraw from Scheduled Courses. Course Rosters are printed which list the students who are registered in each class. The instructor teaches the class and records which of the registered students complete the class and which do not.

Completion Reports and Training Statistics are printed.

Process Name: Maintain Personnel Information

Process Number: 4

Process Notes:

When personnel changes occur, both Group records Student records are updated and a Personnel List is printed.

Appendix C

Detailed Process Model

Table of Contents

Level 1 Data Flow Diagram	150
Design Curriculum	151
Update Curriculum Records	151
Add Curriculum Course	152
Change Curriculum Course	152
Delete Curriculum Course	152
Print Course Catalog	152
Schedule Training	153
Print Needs Form	153
Update Needs	154
Add Need	154
Delete Need	155
List Accumulated Needs	155
Update Schedule	156
Add Scheduled Course	157
Change Scheduled Course	157
Delete Scheduled Course	157
Print Schedules	157

APPENDIX C — DETAILED PROCESS MODLEL

Train Students	158
Update Registration Information	159
Add Enrollment	159
Add Withdrawal Reason	159
Delete Registration	160
Print Roster	160
Add Completion Date	160
Print Completion Report	160
Print Statistics	161
Maintain Personnel Information	162
Maintain Group Information	163
Add Group	163
Change Group	163
Delete Group	164
Maintain Student Information	164
Add Student	165
Change Student	165
Delete Student	165
Print Personnel List	165
Process Allocation	166

Level 1 Data Flow Diagram

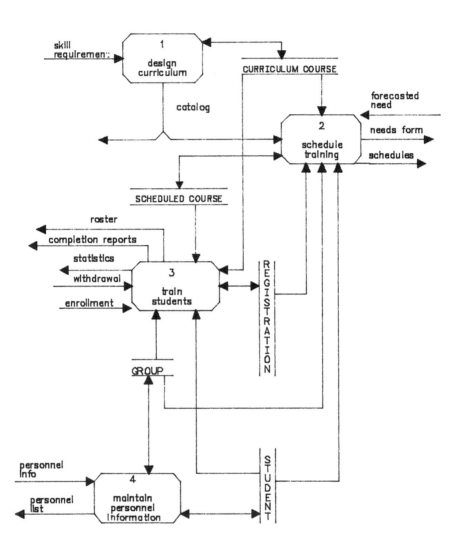

(Manual processes are not included in the Detailed Process Model)

APPENDIX C — DETAILED PROCESS MODEL 151

1. Design Curriculum

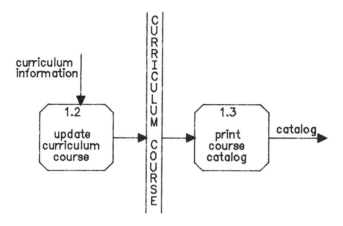

1.2. Update Curriculum Course

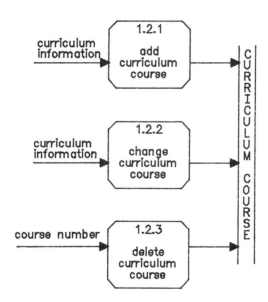

Process Name: Add Curriculum Course

Process Number: 1.2.1

Process Description:

COURSE NAME, COURSE NUMBER, COURSE LEVEL, COURSE DESCRIPTION, and CREDIT HOURS all must be present and valid. COURSE PREREQUISITE if present must be a valid COURSE NUMBER.

Add valid CURRICULUM INFORMATION to the Curriculum Courses. Duplicate COURSE NUMBERS are not valid.

Process Name: Change Curriculum Course

Process Number: 1.2.2

Process Description:

If the COURSE NUMBER exists in the Curriculum Course records, any data item except COURSE NUMBER may be changed.

COURSE NAME, COURSE LEVEL, COURSE DESCRIPTION, and CREDIT HOURS all must be present and valid. COURSE PREREQUISITE if present must be a valid COURSE NUMBER.

Process Name: Delete Curriculum Course

Process Number: 1.2.3

Process Description:

Delete CURRICULUM COURSE INFORMATION for the specified COURSE NUMBER from the Curriculum Course records.

1.3. Print Course Catalog

Process Name: Print Course Catalog

Process Number: 1.3

Process Description:

CURRICULUM COURSE INFORMATION is printed for one COURSE NUMBER or for all COURSE NUMBERS.

2. Schedule Training

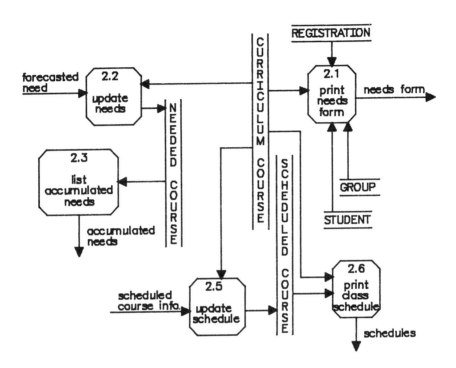

2.1. Print Needs Form

Process Name: Print Needs Form

Process Number: 2.1

Process Description:

Print a form for each Student record. The form contains a list of COURSE NUMBER and COURSE NAME of all Curriculum Courses. COMPLETION DATE is printed for each completed course and a space is provided for the student to indicate which Curriculum Courses he or she wishes to take and the preferred month.

154   THE STRUCTURED SYSTEM LIFE CYCLE: A CASE STUDY

2.2 Update Needs

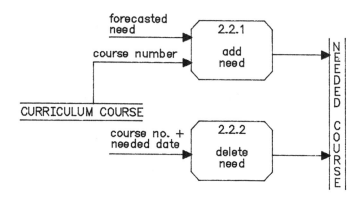

APPENDIX C — DETAILED PROCESS MODEL 155

Process Name: Add Need

Process Number: 2.2.1

Process Description:

COURSE NUMBER must match the COURSE NUMBER of a CURRICULUM COURSE. NEEDED DATE must be present and valid.

If there is no Needed Course record for the COURSE NUMBER and a NEEDED DATE, add a NEEDED COURSE record.

If there is a Needed Course record for the COURSE NUMBER and NEEDED DATE, increase the NUMBER OF NEEDS by one.

Process Name: Delete Need

Process Number: 2.2.2

Process Description:

Delete Needed Course record for the specified COURSE NUMBER and NEEDED DATE.

2.3 List Accumulated Needs

Process Name: List Accumulated Needs

Process Number: 2.3

Process Description:

List Needed Course records by NUMBER OF NEEDS (greatest first), COURSE NUMBER, and NEEDED DATE.

2.5. Update Schedule

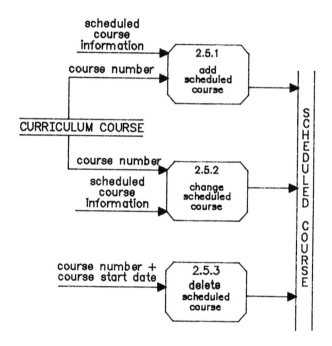

Process Name: Add Scheduled Course

Process Number: 2.5.1

Process Description:

COURSE NUMBER must match the COURSE NUMBER of a Curriculum Course. COURSE COST, COURSE START DATE, COURSE END DATE, COURSE TIME, INSTRUCTOR NAME, and ROOM LOCATION all must be present and valid.

Add valid SCHEDULED COURSE INFORMATION to the Scheduled Course records. Duplicate Scheduled Courses are not valid.

Process Name: Change Scheduled Course

Process Number: 2.5.2

Process Description:

If the COURSE NUMBER and COURSE START DATE exist in the Scheduled Course records, any other data item may be changed.

COURSE COST, COURSE END DATE, COURSE TIME, INSTRUCTOR NAME, and ROOM LOCATION all must be present and valid.

Process Name: Delete Scheduled Course

Process Number: 2.5.3

Process Description:

Delete SCHEDULED COURSE INFORMATION for the specified COURSE NUMBER and COURSE START DATE.

2.6 Print Schedules

Process Name: Print Schedules

Process Number: 2.6

Process Description:

Print SCHEDULED COURSE INFORMATION by SCHEDULED COURSE, by ROOM, and by INSTRUCTOR if the COURSE START DATE is future.

3. Train Students

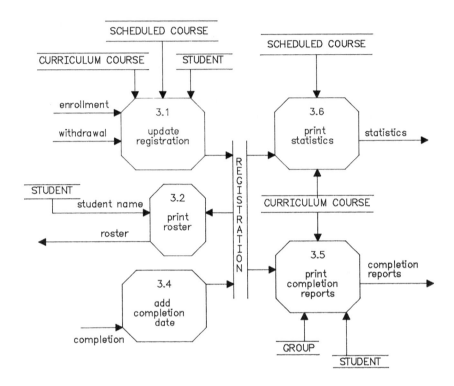

3.1. Update Registration Information

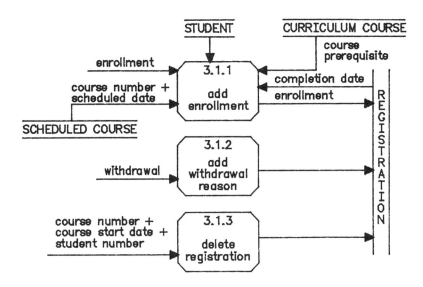

Process Name: Add Enrollment

Process Number: 3.1.1

Process Description:

STUDENT NUMBER must be in the Student records and COURSE NUMBER and COURSE START DATE must identify a Scheduled Course.

The Curriculum Course records are referenced to see if there is a COURSE PREREQUISITE for the course. If there is, then the Registration records are referenced to verify that the student has completed the prerequisite.

Add valid ENROLLMENT to the Registration records. Duplicate ENROLLMENTS are not valid.

Process Name: Add Withdrawal Reason

Process Number: 3.1.2

Process Description:

Add WITHDRAWAL REASON for the specified COURSE NUMBER, COURSE START DATE, and STUDENT NUMBER in the Registration records.

160 THE STRUCTURED SYSTEM LIFE CYCLE: A CASE STUDY

Process Name: Delete Registration

Process Number: 3.1.3

Process Description:

Delete Registration record for the specified COURSE NUMBER, COURSE START DATE, and STUDENT NUMBER.

3.2. Print Roster

Process Name: Print Roster

Process Number: 3.2

By COURSE NUMBER, print STUDENT NAME and COURSE NAME for each Registration record with a future COURSE START DATE.

3.4. Add Completion Date

Process Name: Add Completion Date

Process Number: 3.4

Process Description:

Add COMPLETION DATE for the specified COURSE NUMBER, COURSE START DATE, and STUDENT NUMBER to the Registration records.

3.5. Print Completion Report

Process Name: Print Completion Report

Process Number: 3.5

Process Description:

By GROUP NAME, print STUDENT NAME, COURSE NUMBER, COURSE NAME, COMPLETION DATE, and WITHDRAWAL REASON for each Registration record with a COURSE START DATE during the reporting period.

3.6. Print Statistics

Process Name: Print Statistics

Process Number: 3.6

Process Description:

By COURSE NUMBER and by INSTRUCTOR NAME, print COURSE NAME, COURSE DAYS, number of Registrations with non-blank COMPLETION DATE, and STUDENT DAYS for each Registration record with COURSE. START DATE within the reporting period.

162 THE STRUCTURED SYSTEM LIFE CYCLE: A CASE STUDY

4. Maintain Personnel Information

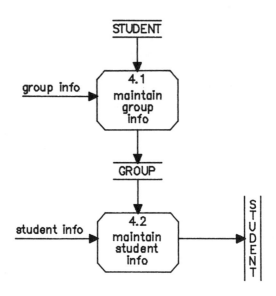

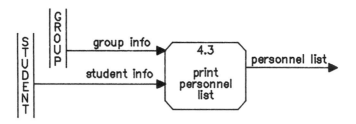

APPENDIX C — DETAILED PROCESS MODEL   163

4.1. Maintain Group Information

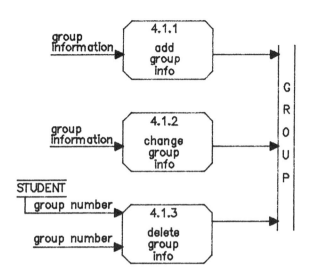

Process Name: Add Group Information

Process Number: 4.1.1

Process Description:

GROUP MANAGER NAME and GROUP NUMBER must be present and valid.

Add valid GROUP INFORMATION to the Group records. Duplicate GROUP NUMBERS are not valid.

Process Name: Change Group Information

Process Number: 4.1.2

Process Description:

GROUP MANAGER NAME must be valid and may be changed for the specified GROUP NUMBER in the Group records.

Process Name: Delete Group Information

Process Number: 4.1.3

Process Description:

Delete GROUP INFORMATION for the specified GROUP NUMBER in the Group records if there are no Student records with that GROUP NUMBER.

4.2. Maintain Student Information

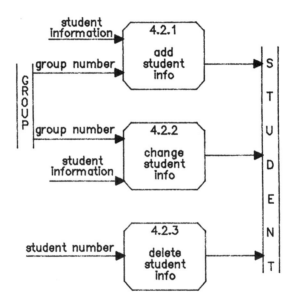

Process Name: Add Student Information

Process Number: 4.2.1

Process Description:

STUDENT NAME, STUDENT NUMBER, and STUDENT LEVEL all must be present and valid. GROUP NUMBER must be a valid Group.

Add valid STUDENT INFORMATION to the Student records. Duplicate STUDENT NUMBERS are not valid.

Process Name: Change Student Information

Process Number: 4.2.2

Process Description:

STUDENT NAME and STUDENT LEVEL must be valid and may be changed for the specified STUDENT NUMBER in the Student records.

Process Name: Delete Student Information

Process Number: 4.2.3

Process Description:

Delete the STUDENT INFORMATION from the Student records for the specified STUDENT NUMBER.

4.3. Print Personnel List

Process Name: Print Personnel List

Process Number: 4.3

Process Description:

Print GROUP INFORMATION and STUDENT INFORMATION in sequence by GROUP NUMBER and STUDENT NUMBER.

Process Allocation

Process 1.1 Change Curriculum and Process 3.3 Teach Class were allocated to a manual processor. The rest of the processes were allocated to a mainframe CPU.

Although not all of the processes had to be on-line to meet the user requirements, the transaction volume is so low there was no reason not to process all the mechanized functions in the same manner. Therefore, the entire system is on-line.

Appendix D

Design Model: Maintain Personnel Information

Table of Contents

Screen Design Layouts	168
Personnel File Menu	168
Add Group	169
Change Group	170
Delete Group	172
Add Student	173
Change Student	174
Delete Student	176
Report Design Layout	177
Personnel List	177
Subsystem Design Structure Chart	178
Program Design	179
Module Design	179
Code Design	182
Route Personnel Transaction	182
Add Group	182
Change Group	183
Delete Group	183
Add Student	184
Change Student	184
Delete Student	185

Screen Design Layouts

Function: 4. ROUTE PERSONNEL TRANSACTION
Screen: Main Menu

```
                *** PERSONNEL FILE MENU ***

                ENTER SELECTION CODE: X

                1 = PRINT PERSONNEL LIST

                2 = ADD STUDENT INFORMATION

                3 = CHANGE STUDENT INFORMATION

                4 = DELETE STUDENT INFORMATION

                5 = ADD GROUP INFORMATION

                6 = CHANGE GROUP INFORMATION

                7 = DELETE GROUP INFORMATION
```

---

   message lines

Function: 4.1.1. ADD GROUP

Screen: 1

*** ADD GROUP INFORMATION ***

GROUP NUMBER: XXX

GROUP MANAGER LAST NAME AND INITIAL: XXXXXXXXXXXXXXX

message lines

Function: 4.1.2. CHANGE GROUP
Screen: 1A

*** CHANGE GROUP INFORMATION ***

GROUP NUMBER: XXX

message lines

APPENDIX D — DESIGN MODEL 171

Function: 4.1.2. CHANGE GROUP

Screen: 1B

*** CHANGE GROUP INFORMATION ***

GROUP NUMBER: XXX

GROUP MANAGER LAST NAME AND INITIAL: XXXXXXXXXXXXXXX

message lines

Function: 4.1.3. DELETE GROUP
Screen: 1

```
                *** DELETE GROUP INFORMATION ***

                       GROUP NUMBER: XXX
```

message lines

APPENDIX D — DESIGN MODEL 173

Function: 4.2.1. ADD STUDENT

Screen: 1

*** ADD STUDENT INFORMATION ***

STUDENT SOCIAL SECURITY NUMBER (no dashes): XXXXXXXXX

STUDENT LAST NAME AND FIRST NAME: XXXXXXXXXXXXXXXX

STUDENT LEVEL: XX

GROUP NUMBER: XXX

message lines

Function: 4.2.2. CHANGE STUDENT
Screen: 1A

*** CHANGE STUDENT INFORMATION ***

STUDENT SOCIAL SECURITY NUMBER (no dashes) : XXXXXXXXX

---
message lines

Function: 4.2.2. CHANGE STUDENT
Screen: 1B

*** CHANGE STUDENT INFORMATION ***

STUDENT SOCIAL SECURITY NUMBER (no dashes): XXXXXXXXX

STUDENT LAST NAME AND FIRST NAME: XXXXXXXXXXXXXXX

STUDENT LEVEL: XX

GROUP NUMBER: XXX

message lines

Function: 4.2.3. DELETE STUDENT
Screen: 1

*** DELETE STUDENT INFORMATION ***

STUDENT SOCIAL SECURITY NUMBER (no dashes): XXXXXXXXX

message lines

Report Design Layouts

Function: 4.3. PRINT PERSONNEL LIST
Report Title: PERSONNEL LIST

mm-dd-yy *** PERSONNEL LIST *** Page XX

GROUP MANAGER NAME	GROUP NUMBER	STUDENT SSN	STUDENT NAME	STUDENT LEVEL
XXXXXXXXXXXXXXX	XXX	XXX-XX-XXXX	XXXXXXXXXXXXXXXXX	XX
XXXXXXXXXXXXXXX	XXX	XXX-XX-XXXX	XXXXXXXXXXXXXXXXX	XX
XXXXXXXXXXXXXXX	XXX	XXX-XX-XXXX	XXXXXXXXXXXXXXXXX	XX

(double space between groups)

178 THE STRUCTURED SYSTEM LIFE CYCLE: A CASE STUDY

Subsystem Design Structure Chart

4. MAINTAIN PERSONNEL INFORMATION

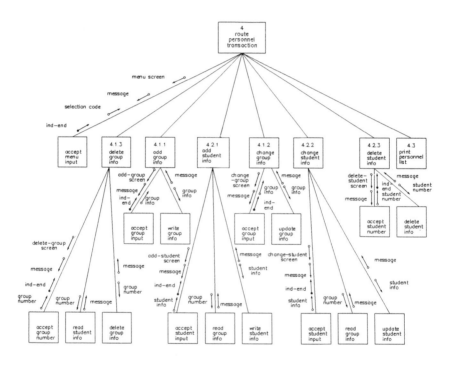

Diagram D.1.

Program Design

The entire subsystem is one program because of the following:

- There does not have to be any time lag between execution of one module and any other.
- There is no reason to have more than one program because of backup and recovery.
- None of the modules need to be run concurrently for any reason.

Module Design

The results of procedural analysis were as follows:

- Accept Menu Input was packaged in the same module as Route Personnel Transaction because they always execute together.
- Route Personnel Transaction is executed only when a different menu option is specified, so no other processes were packaged with Route Personnel Transaction.
- The processes Read Student Information and Read Group Information could be packaged as common modules, but the amount of logic in each of these processes is so small that it was decided to repeat the code instead of sharing it. Also, the program is to be written in a fourth generation language, and file accesses as such will not be coded.
- The file access processes were packaged with their superordinates because most always execute with their superordinates. Another reason for not making any of the file access processes a separate module is that with a fourth generation language the file access logic is not actually coded by the programmer.

180 THE STRUCTURED SYSTEM LIFE CYCLE: A CASE STUDY

- Since only one menu option is executed at a time, no menu option was packaged with any other menu option.

The processes included in each module are shown in the structure chart in Diagram D.2.

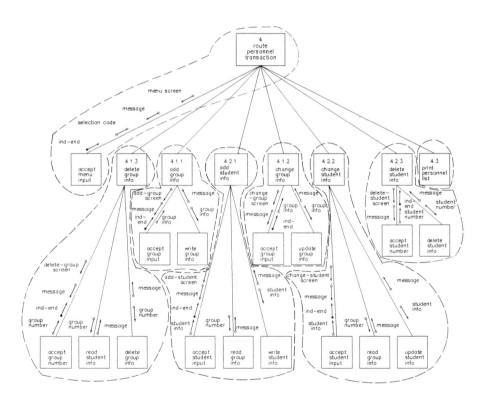

Diagram D.2.

APPENDIX D — DESIGN MODEL   181

The physical module structure of the program is shown in Diagram D.3.

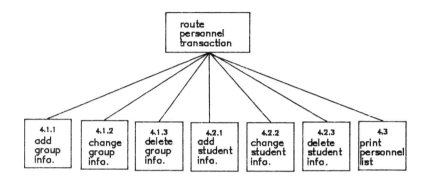

Diagram D.3

Code Design

Route Personnel Transaction

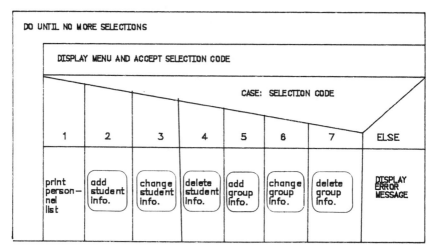

Diagram D.4.

Add Group

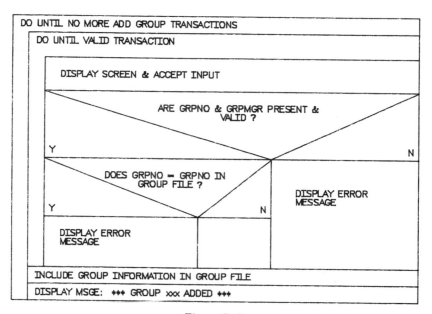

Figure D.5.

APPENDIX D – DESIGN MODEL 183

Change Group

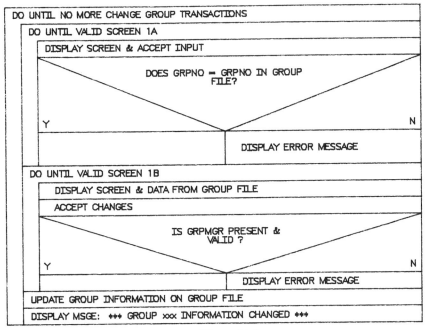

Figure D.6.

Delete Group

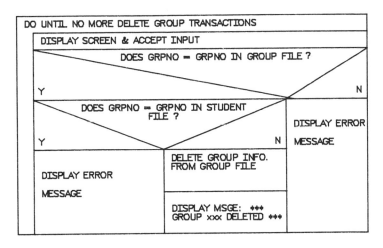

Figure D.7.

Add Student

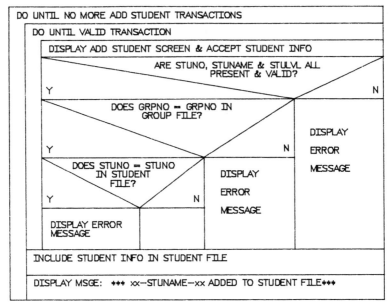

Figure D.8.

Change Student

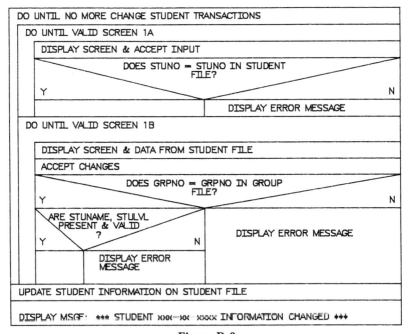

Figure D.9.

Delete Student

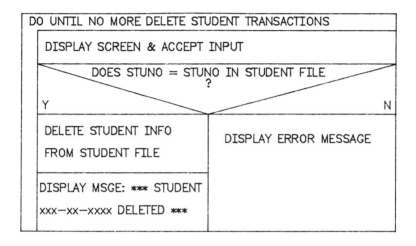

Figure D.10.

Appendix E

Data Dictionary

Table Of Contents

Descriptions	188
Accumulated Needs	188
Completion Date	188
Completion	188
Course Cost	188
Course Days	188
Course Description	188
Course End Date	188
Course Level	188
Course Name	188
Course Number	188
Course Prerequisite	188
Course Start Date	188
Course Time	189
Credit Hours	189
Curriculum Information	189
Enrollment Date	189
Enrollment	189
Group Information	189
Group Manager Name	189
Group Number	189
Ind-End	189
Instructor Name	189
Message	189
Needed Date	189
Number of Needs	189
Personnel Information	189
Room Location	189
Scheduled Course Information	190

Selection Code	190
Student Days	190
Student Information	190
Student Level	190
Student Name	190
Student Number	190
Withdrawal	190
Withdrawal Reason	190
Sample Date Flow Definition	191
Sample Date Element Definition	191

Descriptions

Note: The data dictionary definitions will need to have the alias, data type, length, valid values, and meanings added before coding begins. See sample data element definition and data flow definition.

Accumulated Needs: The data used to determine the courses for which there is the greatest need. Composed of COURSE NUMBER, NEEDED DATE, and NUMBER OF NEEDS.

Completion Date: The date an enrolled student completes all portions of a scheduled course.

Completion: The data required to record the completion of a scheduled course. Composed of COURSE NUMBER, COURSE START DATE, STUDENT NUMBER, and COMPLETION DATE.

Course Cost: The cost in dollars of one occurrence of a scheduled course.

Course Days: The number of days of training provided in a course. Determined by dividing the credit hours by six.

Course Description: The narrative which describes the course content and objectives.

Course End Date: The last day of a scheduled course.

Course Level: The level of the intended course audience.

Course Name: The name used to describe a course provided by the data processing training group.

Course Number: A number which uniquely identifies a course provided by the data processing training group.

Course Prerequisite: The COURSE NUMBER of the prerequisite course.

Course Start Date: The first day of a scheduled course.

APPENDIX E — DATA DICTIONARY

Course Time: The beginning and ending time of the scheduled course. Applies to each day of a multiple-day course.

Credit Hours: The number of hours of training credit given for a course. One day is considered six hours.

Curriculum Information: The data required to describe a course in the data processing curriculum. Composed of COURSE DESCRIPTION, COURSE LEVEL, COURSE NAME, COURSE NUMBER, COURSE PREREQUISTITE, and CREDIT HOURS.

Enrollment Date: The date a student is enrolled in a scheduled course.

Enrollment: The data required to enroll a student in a scheduled course. Composed of COURSE NUMBER, COURSE START DATE, STUDENT NUMBER, and ENROLLMENT DATE.

Group Information: The data required to describe a work group. Composed of GROUP NUMBER and GROUP MANAGER NAME.

Group Manager Name: The name of the manager of a work group.

Group Number: A number which uniquely identifies a work group.

Ind-End: An indicator which is set when the last input screen has been received.

Instructor Name: The name of a person assigned the responsibility for teaching the course.

Message: A response issued by the system indicating either successful execution or an error condition.

Needed Date: The month and year a student anticipates needing to take training.

Number of Needs: The number of students who need the course in a given month.

Personnel Information: The data required to record personnel changes. Composed of GROUP INFORMATION and STUDENT INFORMATION.

Room Location: The room number and building of the classroom.

Scheduled Course Information: The data required to schedule a course. Composed of COURSE END DATE, COURSE NUMBER, COURSE START DATE, COURSE TIME, INSTRUCTOR NAME, and ROOM LOCATION.

Selection Code: A code indicating which menu option has been selected.

Student Days: The number of students multiplied by the course days.

Student Information: The data required to describe a student. Composed of STUDENT LEVEL, STUDENT NAME, STUDENT NUMBER, and GROUP NUMBER.

Student Level: The job level of a student.

Student Name: The name of a student.

Student Number: The social security number of a student.

Withdrawal: The data required for a student to withdraw from a scheduled course. Composed of COURSE NUMBER, COURSE START DATE, STUDENT NUMBER, and WITHDRAWAL REASON.

Withdrawal Reason: The reason why an enrolled student did not complete a course.

Sample Data Flow Definition

DATA FLOW NAME: Group Information

DESCRIPTION:

The data required to describe a work group.

COMPOSITION:

    Group Manager Name
    Group Number

Sample Data Element Definition

DATA ELEMENT NAME: Group Number   ALIAS: GRPNO

DATA TYPE: Integer   LENGTH: 3

DESCRIPTION:

A number which uniquely identifies a work group.

VALID VALUES:*

    000 - 999

* Meaning of the values should also be shown if appropriate.

Appendix F

Answers to Questions and Exercises

Table of Contents

Chapter 1	Questions	193
Chapter 2	Questions	196
Chapter 3	Questions	198
Chapter 4	Questions	199
Chapter 5	Questions	201
Chapter 6	Questions	202
	Exercises	204
Chapter 7	Questions	205
	Exercises	205
Chapter 8	Questions	207
	Exercises	207
Chapter 9	Questions	211
	Exercises	213
Chapter 10	Questions	217
	Exercises	218
Chapter 11	Questions	223
	Exercises	224

Chapter 1
Questions and Answers

1. What would you expect to find in a system development standards manual?

   You would expect to find an explanation of the system development life cycle, including the phases; what is to be done in each phase; what deliverable documents are to be produced and what each is to contain; what methods and forms are to be used; and the review and approval process for each deliverable document.

2. What is the difference between software development and system development?

   System development includes both manual and mechanized functions; software development includes just the mechanized functions.

3. What system development phase is Alan skipping? Should he be? Why?

   Alan is skipping the Feasibility Study phase. The purpose of a Feasibility Study is to determine whether a project is economically, operationally, and technically feasible to do. It includes a cost/benefit study for several different alternative ways to solve a problem. If the project is obviously operationally and technically feasible, and the cost is not great or the project has to be done regardless of the cost, there is no problem in skipping the feasibility study. However, Alan does not know whether or not a Feasibilty Study was done. He did not ask, and the manager did not say. He should have asked because perhaps some of the data gathering has been done already.

4. What is the purpose of the Requirements Definition phase?

   The purpose of the Requirements Definition phase is to ensure that the developers and the customer have the same understanding as to what the problem is and how it is going to be solved. The level of detail should be only what is needed to accomplish that goal. It is important that all users are identified and that the needs of all are addressed.

194   THE STRUCTURED SYSTEM LIFE CYCLE: A CASE STUDY

5. Why is it important to formally define the scope of a project?

It is important to formally define the scope of the project because often the scope is not as apparent as it seems. Many projects are perceived as unsuccessful because the team omitted something that was important to the customer or was over time and over budget because they were developing something that was not important to the customer.

6. What is meant by software development tools?

Software development tools are products that aid the analysts and programmers in their work. These include text editors, code generators, mechanized libraries, graphics such as data flow diagrams, structure charts, and entity-relationship diagrams. Or they may be products that assist in designing databases.

7. When is true prototyping justified in business software development?

True prototyping is developing a real working model of the system. It rarely can be justified in business software development. In fact, it can be dangerous. Since the prototype is thrown together as quickly as possible without much thought given to design, it is very similar to systems built twenty years ago before the use of structured analysis and design. Because it does work, the pressure to keep it and never develop a system that can be maintained is almost impossible to counteract.

If the prototype is a throw away system, the cost of developing two systems seldom can be justified in a business environment. Some people use the argument that a prototype is useful when there is no other way to determine the system requirements. While this argument is valid in theory, examples of this occurrence in the business world are rare.

The type of prototyping that is very useful in the business environment is where screen interactions and reports are simulated.

8. What do you think about the manager's comment regarding using prototyping and skipping some phases?

APPENDIX F — ANSWERS TO QUESTIONS & EXERCISES 195

The manager's comment about prototyping and skipping some phases comes from reading trade literature without really understanding it. Prototyping is simply a way of defining requirements faster and more accurately; it does not eliminate any steps.

9. What are some tools you could use for prototyping?

Fourth generation languages can be used for prototyping (simulation), as well as any other means of quickly designing screen interactions and report layouts without having the real databases and edits working.

10. What are some of the questions that Alan should ask the training coordinator in his first interview?

The questions should draw out the business objectives, the events the system must respond to, the names and location of all users, the system inputs and outputs, and any hidden agendas the users may have.

11. What is meant by the term *user*?

The term *user* is used to designate any area or other mechanized system that is a source of system inputs or the recipient of system outputs. The term *customer* now often is used to designate the system owner for whom the system is being constructed.

12. According to *Fundamental Concepts of Information Modeling*, upon whose work is information modeling based? Who is Ted Codd? How do his data analysis techniques differ from information modeling?

Information modeling is based on the work of Peter Chen at UCLA in 1976. Ted Codd worked for IBM and developed much of the relational database theory. Ted Codd's techniques for data analysis primarily are mathematical, whereas Peter Chen's are based more on use of the data in the business environment.

Chapter 2
Questions and Answers

1. Why do you think that Alan did not diagram the registrar as a user of the course schedule?

    The registrar performs part of the training function and therefore is inside the system.

2. Why do you think that Alan did not diagram the training coordinator as the source of the curriculum?

    The training coordinator also performs part of the training function and therefore is inside the system.

3. If the system did not exist already as a manual function, what would have to be done differently?

    The users would have to define the functions first. The analyst's job will be more difficult because he will be trying to mechanize functions that have not been tried manually. The potential for an unusable system is therefore greater.

4. If the system already were partly mechanized, what additional information would Alan need to gather?

    He also would want to look at record layouts and data dictionary definitions if there are any.

5. Write the problem definition.

    The DP training area keeps records about requested training, curriculum courses, scheduled courses, enrollments, completions, training resources, and students. It produces reports that provide information about enrollments, forecasted needs, completions, scheduled courses and resource usage. In addition, it produces a course catalog.

    The records are being kept manually on the word processor. As a result, they are not consistent, updating them is very time consuming, and the paper work is becoming unmanageable.

APPENDIX F — ANSWERS TO QUESTIONS & EXERCISES 197

6. What are the project objectives? (Objectives must be measurable.)

    One business objective is to eliminate the need for two clerical positions by providing complete and consistent training records and by minimizing the time and effort required to update them.

    A second business objective is to save 25 percent of the registrar's time by providing quick and flexible reporting.

7. List the user needs defined so far.

    Terri has mentioned the following:

    - Record no-shows by date and indicate the reason.
    - Report completions and no-shows by time period by individual, group, course, and instructor.
    - Produce course rosters by month.
    - Produce the forecasted needs turn-around document.
    - Mechanically generate course schedule.
    - Provide the capability for students to enter their own enrollments on-line.
    - Validate prerequisites at enrollment time.
    - Mechanically produce the course catalog.
    - Generate training statistics regarding student days and room usage.

8. List the events to which the system must respond.

    - Job skill requirements change.
    - Student foresees training need.
    - Student decides to take a course.
    - Personnel change occurs.

9. Are there any constraints that we are aware of so far?

   Alan seems to be the only person assigned to work on the project, so there may be a resource constraint.

   If the students are to do their own enrollments, there may be a constraint that the system must use the type of terminals that the students already have.

Chapter 3
Questions and Answers

1. What is meant by business data? Data item?

   By business data, we mean data that the customer uses and knows about. We do not want to include data elements that are used only internally in programs. We may want to add some of that data later, but not now.

   By data item, we mean the smallest piece of information that is used by itself (sometimes referred to as data token). This is determined by its use in the system being analyzed. For example, one system might treat a person's name as one single piece of information, while another system might subdivide it into last name and first name.

2. How would Alan find the data items if this were an entirely new system that had not existed in a manual procedure?

   He would have to determine what data is required solely by interviewing the users. This process is much more time consuming. In his system, he only needs to interview to find any additional information the mechanized system will need.

3. What is meant by derived data? Do you want it in the data dictionary? Why?

   Derived data is information that can be obtained from some other data item or items.

We do not store derived data in the data dictionary at this point. An example would be CREDIT DAYS. Since we know CREDIT HOURS, we know we can calculate CREDIT DAYS if we want to.

At this point, we want to keep the data dictionary as simple and uncluttered as possible.

4. What other things will Alan be putting into the data dictionary as he develops the system?

The data dictionary contains more than just data items. It also contains, among many other possible things, process descriptions, data store descriptions, data flow definitions, object definitions, and relationship definitions.

Chapter 4
Questions and Answers

1. What is an object?

   An object is a person, place, thing, or event about which an organization stores or wishes to store information.

2. What is an attibute?

   An attribute is the data that is stored about an object which describes or identifies the object.

3. Define the different types of objects and explain how you know which type an object is.

   Independent object, subtype and supertype objects, characteristic object, and associative object.

   An independent object is one that does not need any related objects in order to exist.

   A subtype object is a dependent object which further describes a particular role played by an object. It inherits the properties of its supertype.

A supertype object is one which may play several different roles and has a subtype object for each role.

An associative object is an object that may exist solely to associate other objects, often a formal document such as a contract.

A characteristic object is an object that in turn defines, or characterizes, another object.

4. What is another name for an object-relationship diagram?

Another name for an object-relationship diagram is an entity-relationship diagram.

5. If Alan finds that a subtype object has no attributes that uniquely apply to it, what should he do? What if the only attribute of a subtype object is its number or name?

A subtype object that has no attributes should be collapsed into its supertype.

A subtype object whose only attribute is its number or name also should be collapsed into its supertype.

6. Was Alan smart to continue work on the information model before he could talk to Terri about it?

In this case, he probably was smart to continue work on the information model because he had a fairly good understanding of how the system had to work. If this had not been the case, though, he could have been wasting his time.

7. How could Alan have avoided wasting time while he waited to talk to Terri?

Alan could have avoided wasting time waiting to talk to Terri by anticipating when he was going to have to talk to her and planning ahead for an appointment.

8. Do you know of any techniques, other than prototyping, for speeding up requirements definition?

APPENDIX F — ANSWERS TO QUESTIONS & EXERCISES 201

There are several techniques in use for defining requirements in a group session. For example, IBM has a method called Joint Application Design; Boeing Computer Services has a method called Consensus.

Chapter 5
Questions and Answers

1. If Alan were more experienced, what question would he have asked Terri when she kept talking about the room and instructor data?

   He would have asked her what she thought she would do with this information in the future. Obviously, she has some future plans that are not coming out in the conversation. Knowing what the next system enhancement is likely to be could influence the way the system is designed.

2. Is it possible for an object to have different attributes, depending on the application?

   Definitely. That is one of the features of information modeling. You only model the information as your application views it.

3. What determines whether or not an object is included in the information model for a particular system?

   An object only is included in the information model if information is or will be stored about it in the particular system being modeled.

4. What test can be used to tell if an object is associative?

   An associative object cannot exist by itself. If any of the objects it associates is removed, the associative object will not have meaning.

5. Alan eliminated the objects ROOM and INSTRUCTOR because he had only one data item attributed to each. Under what condition would he *not* have eliminated an object that had only one attri-

bute?

He would not have eliminated an object which had only one attribute if that attribute were not its unique identifier. Every object must have a unique identifer to distinguish it from other objects of the same type. If it does not, one will have to be constructed for it. Then there would be two attributes and the object should not be eliminated from the model.

6. What do we know about the relationship between objects and data stores?

Objects become the data stores in the process model.

7. Why do we attribute a data item to only one object?

We attribute a data item to only one object to eliminate data redundancy in the database.

8. Why do we define the relationships so precisely?

Precisely defined relationships will be needed for database design.

9. What approaches can be used to determine the data items which describe the objects in a system?

One approach is to list all the data items on all the documents and then attribute each to an object. Another approach is to have the users list what they need to know about each object and then use any existing documents for verification.

Chapter 6
Questions and Answers

1. Do you detect anything that Alan omitted?

Alan did not label the data flows to and from the data stores; however, this is permissible unless it is not obvious what the data are.

2. Do you think the level of detail on Alan's data flow diagram is appropriate for a Level 1 Data Flow Diagram? Why?

APPENDIX F — ANSWERS TO QUESTIONS & EXERCISES 203

The level of detail on Alan's data flow diagram is appropriate because there are four processes. This is within the maximum of nine processes per diagram.

3. Alan has shown some processes that may not be mechanized or that are a mix of manual and mechanized. Should he?

    At this point in the analysis, all processes included within the scope of the system should be shown. The processes have not yet been allocated to manual or mechanized processors.

4. Can an event have more than one system input associated with it? Can an event have system outputs associated with it?

    An event can have more than one system input associated with it and also may have system outputs associated with it.

5. The data store Needed Course does not appear on the diagram. Why?

    The data store Needed Course is inside the process Schedule Training. It will appear when that process is exploded.

Exercises and Solutions

1. Draw a context diagram to depict the following transaction processing system:

   Transactions are large or small, red or blue, and round, square, or oblong. All transactions must be edited for size, shape, and color. An error report showing which errors were found in each rejected transaction is returned to the user.

   An amount is calculated for large transactions. This amount is the quantity in the transaction multiplied by a calculation factor stored in a table. All transactions, including the calculated amount, are written on a file. Red square and blue round transactions are printed on an exception report, which goes back to the user.

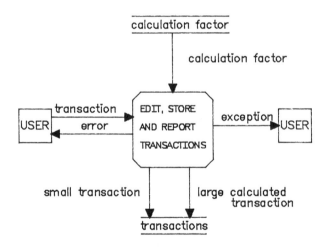

Diagram 1.1. Context Diagram.

APPENDIX F — ANSWERS TO QUESTIONS & EXERCISES 205

Chapter 7
Questions and Answers

1. Is Alan smart to proceed with the detailed analysis and module design before the screen formats are defined? Why?

   In this case, yes. He knows enough about the system to be fairly confident that any additional requirements which might be discovered during design of the screen and report formats will not significantly affect the design of the processing. He will be able to complete the system faster by doing these activities concurrently.

   If he finds much additional information being discovered, however, he might be wise to stop doing the detailed analysis and module design until the screens are designed.

2. What tasks will Alan be doing in detailed analysis?

   In detailed analysis, Alan will be fully exploding the data flow diagrams, adding data flow and data element descriptions to the data dictionary, writing mini-specs, and describing the data stores.

3. What did Alan forget to document? Why do you think he forgot it?

   Alan forgot to document the course cost that Terri needed to have added.

   He probably forgot because he was so concerned with the due date and because someone else was designing the layouts where course cost appears.

Exercises and Solutions

Using the exercise you were given in Chapter 6, do the following:

1. List each data element in each system input and output.

   Transaction — Input:
   size, color, shape, quantity

   Calculation Factor — Input:

calculation factor

Large Calculated Transaction — Output:
size, color, shape, quantity, amount

Small Transaction — Output:
size, color, shape, quantity

Exception — Output:
size, color, shape, quantity, amount

Error — Output:
size, color, shape, quantity, error message

2. List all the different transactions and assign an assumed volume to each. An example is a large, red, round transaction.

   One assumption would be an equal number of red and blue, an equal number of large and small, and an equal number of round, square and oblong. Assuming 120 transactions are entered, each transaction type would have a volume as follows:

Transaction Type	Volume
red round small	10
red round large	10
red square small	10
red square large	10
red oblong small	10
red oblong large	10
blue round small	10
blue round large	10
blue square small	10
blue square large	10
blue oblong small	10
blue oblong large	10

Chapter 8
Questions and Answers

1. What is a quick way to tell what level data flow diagram you are looking at?

   By the number of digits in the process number. X. indicates Level 1; X.X. indicates Level 2; X.X.X. indicates Level 3.

2. Why is the detailed process model documentation not part of the Requirements Definition document?

   The purpose of the Requirements Definition document is to ensure that what the customer thinks is going to be developed is the same as what the analyst is planning to develop. The detailed process model will add very little to that understanding, too much detail will make it more difficult for the customer to assimilate the required information, and the analyst could have wasted time defining details for the wrong problem.

Exercises and Solutions

Using your answers to the exercises in Chapters 6 and 7, do the following exercises:

1. Explode the Context Diagram as far as possible, and evaluate the data flow diagrams according to the following rules.

   - All system inputs and their sources and all system outputs and their recipients should show on the context diagram.
   - As the diagrams are exploded, no data flows or data stores should be dropped, and the only ones that should be added are those which were totally inside the higher level process.
   - No data should flow to a process where it is not used.
   - Process names should consist of action verbs and objects.
   - No process should only rearrange data.

   The class solution as shown in Diagram 1.1. adheres to the above criteria, and there may be other acceptable solutions.

208 THE STRUCTURED SYSTEM LIFE CYCLE: A CASE STUDY

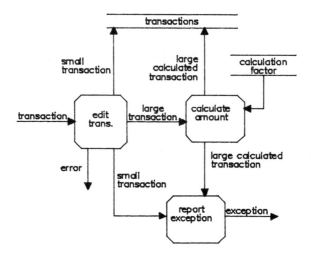

Diagram 1.1. Exploded Data Flow Diagram

It is possible to further explode the process Edit Transaction as in Diagram 1.2.

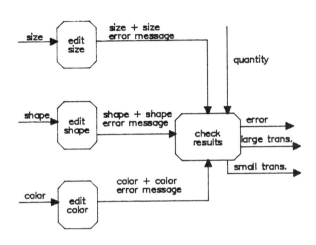

Diagram 1.2. Level 2 Data Flow Diagram.

Whether or not it is advisable to explode to this level of detail would depend on how large the mini-spec would be to describe the unexploded process, whether further explosion adds or detracts from the understanding of the problem, and whether the exploded diagram looks more complicated than the unexploded diagram did. In this case, it probably is not preferable to explode the process further.

2. Write mini-specs and data dictionary definitions (You may need to use your imagination.)

Process Name: Edit Transaction
Process Description:

If SHAPE is not valid then
    Assemble SHAPE ERROR MESSAGE

If SIZE is not valid then
    Assemble SIZE ERROR MESSAGE

If COLOR is not valid then
    Assemble COLOR ERROR MESSAGE

If SHAPE or SIZE or COLOR was not valid then
    Reject Transaction with ERROR MESSAGE

Process Name: Calculate Large Amount
Process Description:

If QUANTITY is not numeric then
    AMOUNT = zero

Otherwise if QUANTITY = zero then
    AMOUNT = zero

Otherwise AMOUNT = QUANTITY X CALCULATION FACTOR

Process Name: Report Exceptions
Process Description:

If COLOR is red or blue and
    SHAPE is square or round then
        Report Exception

APPENDIX F — ANSWERS TO QUESTIONS & EXERCISES 211

Chapter 9
Questions and Answers

1. At what point in the development life cycle does it matter who or what will perform a particular function?

    Process allocation, which is between analysis and subsystem design.

2. How does the use of a fourth generation language affect design?

    In process allocation, consideration must be given to the question of which processes are best suited to be coded in a fourth generation language versus a language such as COBOL.

    In subsystem design, processes allocated to a fourth generation language do not need report formatting, file access, or table handling processes added because the programmer does not have to write procedural code for those processes.

    In module and program design, fourth generation language processes might be packaged differently because of size or execution considerations.

    In code design, many of the procedural instructions do not have to be coded because the fourth generation language takes care of them.

3. How is design affected by the analysis phase output?

    One of the major design principles is to design modules which are functionally cohesive. When design is done using expanded data flow diagrams, the processes in the design are by definition functionally cohesive. This greatly expedites the design phase.

    Another major design principle is to design modules which have minimal coupling. Coupling is minimized when data flow diagrams are drawn following the principle of "starve the bubbles," meaning that no data flows to a process where it is not used. Again, this greatly expedites the design phase.

Data stores that have been determined through information modeling already are normalized, greatly expediting database design.

The business data are already in the data dictionary, so all that needs to be added during design is any additional data required for implementation.

4. Why would you want more than one subsystem in a system?

Smaller systems are much quicker to design, test, and maintain than large systems. For example, five subsystems will take less time to design, test, and maintain than one system containing the same processes *if they are truly separate subsystems.*

5. What type of processes are added during design?

Processes that perform file access, data formatting, table handling, and execution control.

6. Could Alan have coded directly from his data flow diagrams rather than doing design? Why?

The personnel subsystem is so simple to design that he possibly could have coded directly from the data flow diagrams. There are some dangers involved in doing that, however.

If he had coded from the *exploded* data flow diagrams, he undoubtedly would have coded a control process for the group transactions and a control process for the student transactions — the Level 2 Data Flow Diagrams. This would have created an additional unnecessary level of control.

If he had coded from the *expanded* dataflow diagram, he probably would not have had the extra control level, but the module hierarchy would not be documented. Coding a good hierarchical design from an expanded data flow diagram for a large subsystem could be very difficult. For example, without the structure chart, it is difficult to identify common logic which can be invoked from more than one module.

Procedural analysis is difficult to do using data flow diagrams. The decisions, loops, and flow of control are needed to determine optimal packaging of the processes into modules and programs.

It may be possible to produce well-designed code directly from data flow diagrams, but not advisable. If the subsystem is so small that a structure chart is not absolutely necessary, the amount of time required to produce one will be negligible. There is great merit in adhering to a consistent set of procedures regardless of the size of the subsystem. That way, everyone on the development team and everyone who will maintain the system knows exactly what they have to work with.

Exercises and Solutions

Using the results of the exercises in Chapter 8, do the following exercises:

1. Indicate the input and output streams and the central transform on the expanded data flow diagram.

   Since we decided that the Level 1 Data Flow Diagram did not need to be exploded further, the expanded data flow diagram is the same as the Level 1, as in Diagram 1.1.

214 THE STRUCTURED SYSTEM LIFE CYCLE: A CASE STUDY

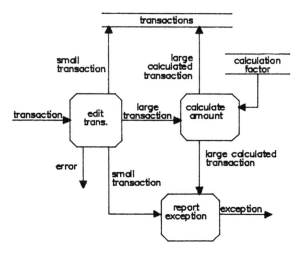

Diagram 1.1. Exploded Data Flow Diagram

If we had decided that Edit Transaction should have been exploded another level, its Level 2 Data Flow Diagram would have replaced its Level 1 Data Flow Diagram on the expanded data flow diagram.

The input stream contains the processes that prepare the raw input data for use by the central transform. Edit Transaction is the input process.

The output stream contains the processes that take the output from the central transform and make it suitable for use by the recipients. Report Exceptions is the output process.

That leaves Calculate Amount as the central transform.

2. Draw the structure chart.

APPENDIX F — ANSWERS TO QUESTIONS & EXERCISES

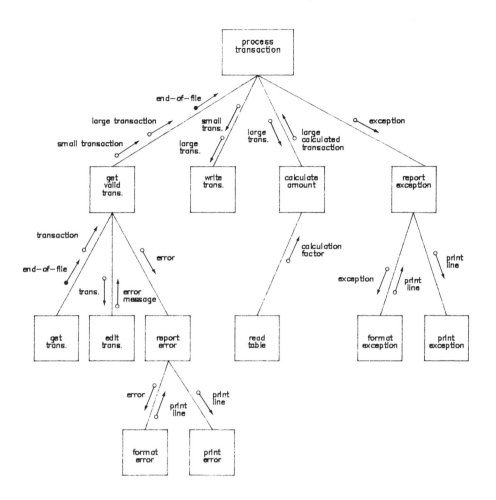

Diagram 2.1. Structure Chart.

3. Which processes would not need to be added if the system were going to be coded in a fourth generation language?

   Get Input, Format Exception, Print Exception, Format Error, Print Error, and Read Table. The structure would look like this Diagram 3.1.

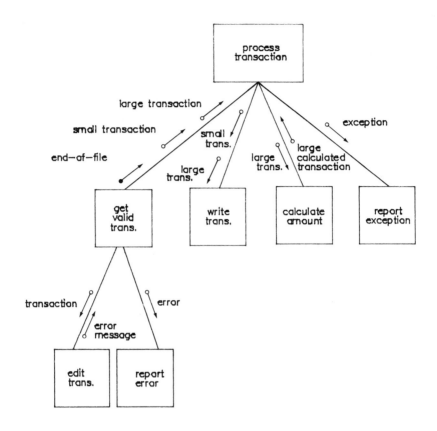

Diagram 3.1. Structure for Fourth Generation Language.

Chapter 10
Questions and Answers

1. What is packaging?

    Packaging is the process of grouping processes into modules and programs.

2. List the steps in procedural analysis.

    The purpose of procedural analysis is to group processes into modules which will execute efficiently and produce modules with a reasonable amount of code. The steps are as follows:

    - Determine the approximate amount of code required in each process.
    - Determine the number of times each process will be executed in a given period.
    - Indicate which processes are part of a loop.
    - Indicate which processes have more than one superordinate.
    - Indicate which processes always execute whenever some other process executes.
    - Put one-time or seldomly executed processes into separate modules.
    - Put processes which have more than one superordinate in a separate module unless the amount of code will be so small that it is not practical.
    - Do not separate processes that are part of the same loop unless the amount of code in the module will be very large.
    - Do not separate processes that always execute together unless the amount of code in the module will be unacceptably large.

3. What criteria do you use to group modules into programs?

The purpose of grouping modules into programs is to provide for back-up and recovery, security, and concurrency of execution. The criteria are the following:

- Make the run time of each program small enough so that rerunning it would not unduly impact the operations schedule.
- Separate the modules according to who has authorized access to them.
- If one group of modules is not dependent on the output of another group of modules, allow them to executed concurrently to minimize elapsed run time.

4. At what point in the development life cycle do you consider the efficiency of the code?

Assuming that the language is coded intelligently, efficiency of the code is considered during code design and coding if it does not adversely affect maintainability. After the programs are coded and running, they should be analyzed to see where the bottlenecks are and only the code that has a significant impact on efficiency should be "tuned." It is usually not productive to "tune" code that executes infrequently. On the other hand, it may be worthwhile to "tune" code in daily runs which handle high volumes of input. It is always good to remember that the time it takes for a programmer to make a maintenance change probably costs more than the extra time that the computer uses because of inefficent code.

Exercises and Solutions

1. Use procedural analysis on the structure chart that you produced in the exercises in Chapter 9. Group the processes into modules and explain your rationale. Assume that the system is at least three modules and that 20 of the 120 transactions will reject because of error.

APPENDIX F — ANSWERS TO QUESTIONS & EXERCISES

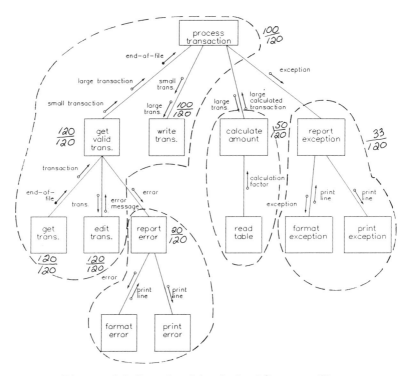

Diagram 1.1. Procedural Analysis of Structure Chart.

Report Error happens infrequently, but when it does occur Format Error and Print Error also occur. These three will be packaged into one module. Report Exception occurs a third of the time, so it, along with Format Exception and Print Exception, will be packaged into one module.

Calculate Amount occurs only on large transactions, which is roughly 50 percent of the time, and always invokes Read Table. They are logical candidates for being together in a separate module.

Get Transaction and Edit Transaction occur for every transaction, and Process Transaction and Write Transaction occur for all valid transactions. All of these should be packaged into the same module.

None of the resulting modules need to be subdivided because of being too large.

2. Draw an N/S chart for each module. Refer to the process descriptions in Chapter 8, Exercise 2.

220 THE STRUCTURED SYSTEM LIFE CYCLE: A CASE STUDY

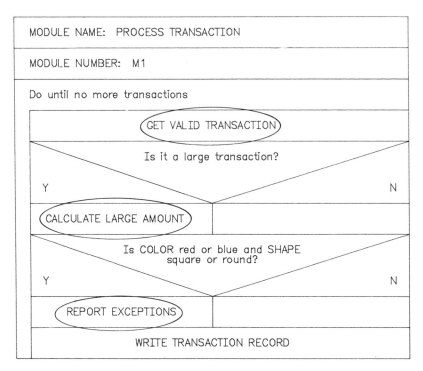

Diagram 2.1. Nassi-Shneiderman Diagram: Process Transaction.

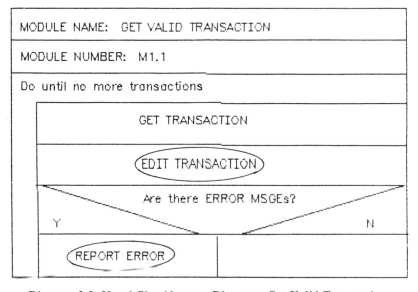

Diagram 2.2. Nassi-Shneiderman Diagram: Get Valid Transaction.

APPENDIX F — ANSWERS TO QUESTIONS & EXERCISES 221

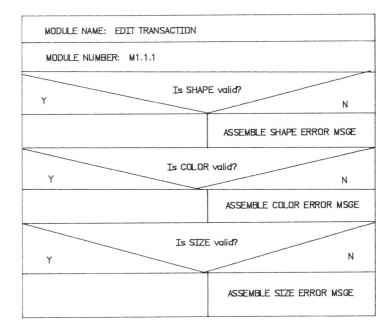

Diagram 2.3. Nassi-Shneiderman Diagram: Edit Transaction.

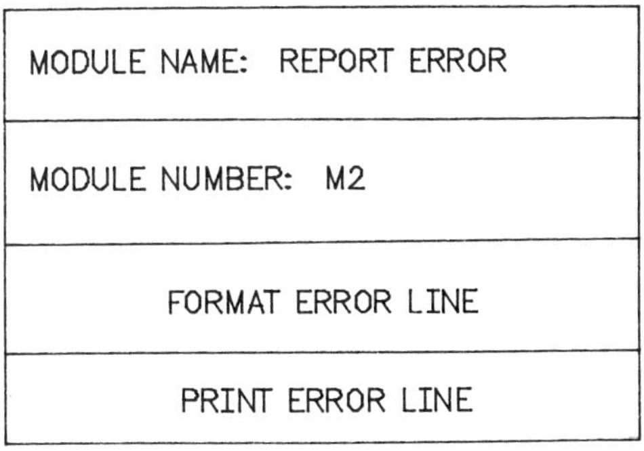

Diagram 2.4. Nassi-Shneiderman Diagram: Report Error.

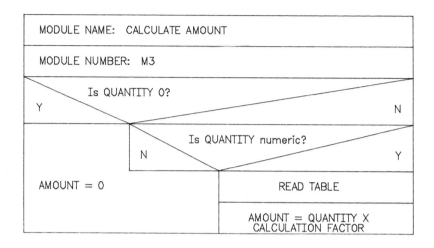

Diagram 2.5. Nassi-Shneiderman Diagram: Calculate Amount.

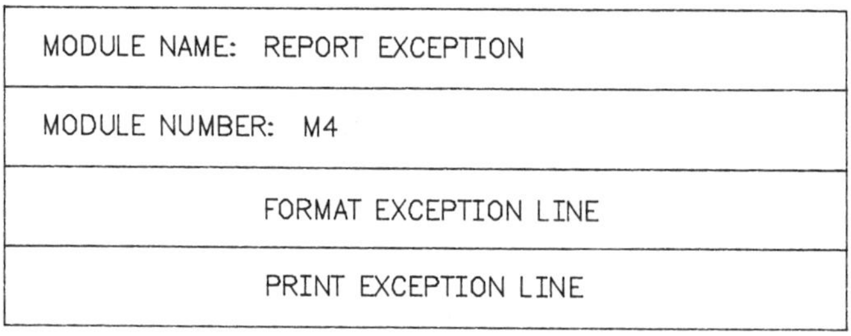

```
┌─────────────────────────────────────────────┐
│ MODULE NAME:  REPORT EXCEPTION              │
├─────────────────────────────────────────────┤
│ MODULE NUMBER:  M4                          │
├─────────────────────────────────────────────┤
│         FORMAT EXCEPTION LINE               │
├─────────────────────────────────────────────┤
│         PRINT EXCEPTION LINE                │
└─────────────────────────────────────────────┘
```

Diagram 2.6. Nassi-Shneiderman Diagram: Report Exception.

Chapter 11
Questions and Answers

1. What is the general rule about the steps to be followed for maintenance changes?

    The general rule is to start as far back in the development life cycle as something changed and modify every affected piece of documentation.

2. How does structured development impact the maintenance effort?

    Information modeling impacts the maintenance effort by minimizing changes to the system files and databases. Files and databases that reflect the way the company uses the data will have fewer changes. Minimizing data redundancy means fewer files to modify when a change does occur.

    Structured analysis impacts the maintenance effort by minimizing the number of changes requested. Because user requirements have been defined more rigorously, more of the user needs will have been met initially.

    Structured design impacts maintenance by minimizing the "ripple effect" of changes. When each process contains one and only one function and these functions are as independent as possible, changes are isolated to small areas of the system and can be made more quickly and accurately. Testing is simplified. Structure charts provide blueprints of the system architecture so that the area to be changed can be identified quickly.

    Structured Coding impacts maintenance because the code is less error prone and is easier to change.

Exercises and Solutions

1. In the problem that you have used in these exercises, explain what would need to be changed under the following circumstances:

    a. If green transactions were now allowed?

   The requirements have changed. A determination would have to be made about whether or not the new transactions were exceptions.

   The data dictionary will have to be changed to add another valid value for COLOR.

   The Edit Transaction process description will need to be changed if the new transactions are exceptions.

   The volume of the new transactions would need to be determined. This potentially could change the packaging of the modules.

    b. If the CALCULATION FACTORS in the table were changed?

   Perhaps nothing. If the change involved changing the format of the calculation factor table, then perhaps the code would need to be changed.

    c. If LARGE TRANSACTIONS were to be written on a separate file from SMALL TRANSACTIONS?

   The data flow diagrams would need to be changed to show the new data store.

   The data elements would have to be listed for the new output.

   The structure chart would need to be modified.

   N/S charts and code would have to be changed.

Glossary of Terms

ad-hoc reports: special or one-time reports; not produced on a regular basis.

analysis: the examination of a problem to distinguish its component parts, both separately and in their relation to the whole.

application: the problem to which a computer is applied to operate in a unified manner for performing specified business functions.

associative object: an object which exists solely to associate other objects; often a formal document.

attribute: data item which describes an object.

backup: the process of making a copy of data, usually for use in the event that the original is destroyed.

business data: data required by the business, regardless of who or what performs the functions.

case: a set of actions whose performance is dependent on a corresponding set of conditions.

central transform: the major data-transforming function of a system.

characteristic object: an object which describes, or characterizes, another object.

chargeback: costs incurred by one area of a corporation and recorded as expenses for a different area.

cohesion: a measure of the strength of association of processing activities.

constraint: any restriction on development of a system which could adversely impact meeting the system objectives.

control function: a process whose sole function is to control the execution of other processes.

context diagram: the top-level diagram of a leveled set of data flow diagrams.

conversion: the activities required to put a newly developed system into production for the first time.

coupling: a measure of the interconnection between processes.

data: information of known composition.

database: a collection of interrelated data stored together.

data flow: a pipeline along which data is passed; data in motion.

data flow diagram: a graphic tool for depicting the partitioning of a system into a network of activities.

data item: the smallest unit of information used in the system; data element.

data store: a repository for data; data at rest.

debug: to remove defects.

decision chart: a tabular form of a decision tree.

decision tree: a graphic tool for portraying a hierarchy of independent conditions and the activities resulting from each valid combination of conditions.

derived data: information which can be determined from other data.

design: to fashion according to plan; the synthesizing of a network of logical elements to perform a specific function.

development: step-by-step construction.

development center: a function within the data processing area that provides tools, training, and assistance in system development.

documentation: the production of documents substantiating knowledge or decisions.

efficiency: using few resources.

entity: object.

executed: performed; accomplished.

expanded data flow diagram: a network of all the lowest level processes of the data flow diagrams.

exploded data flow diagram: a further partitioning of a data flow diagram.

factored: partitioned; each separable process identified.

feasibility: practicality; capability of being dealt with successfully.

file: an organized collection of data in the same location.

file access: a process by which data in a file is stored, retrieved, or changed.

format: organizing data into a specific configuration.

fourth-generation language: a coding language that allows the programmer to specify in English words what is to be accomplished rather than how it is to be performed.

frequency: how often a process is performed in relation to the total number of inputs.

function: a business activity.

implementation: the activities of coding, testing, and debugging a system.

information modeling: identifying objects and relationships between the objects, and attributing the data items wich describe the objects and relationships.

information requirements: the data items, source or recipient, frequency, volume, security, retention, and purpose of system inputs or outputs.

interactive system: a system in which commands from a terminal operated by the user call forth a response from a system or program.

involved: activated.

logical: free of the constraints or characteristics of any particular implementation.

loop: the repeated execution of a series of instructions.

maintenance: changes to existing systems due to errors, changing requirements, or changing environment.

mini-specs: process descriptions for each process on the lowest level data flow diagrams.

model: a representation of some real system which predicts the behavior of that system.

module: a collection of program statements that are implemented together; for example, the COBOL statements that comprise a compilable unit, or the FOCUS statements contained in a FOCEXEC.

Nassi-Shneiderman chart: a diagram which shows the control and processing sequence of modules using three constructs: sequence, decision, and iteration.

object: a person, place, thing, form, or event about which a corporation stores or wishes to store data.

on-line system: a system in which input data is transmitted directly into the computer from the point of origin.

packaging: grouping processes into modules and programs.

PC: a personal computer.

phase: a group of logically related system development steps, for the purpose of more easily managing the development process.

physical implementation: development of procedures for manual processes or code in a specific language to be executed on a specific computer for mechanized processes.

procedural analysis: the process of analyzing the procedural characteristics of processes to be mechanized; frequency of execution, sequence of execution, volume of input, etc.

process: an activity to be performed to accomplish a business function.

process allocation: determining by whom or what a particular process will be performed.

process description: the narrative stating that the policy by which input data is transformed into output data.

program: a group of program statements invoked by the operating system; it may include one or many modules.

project: a series of tasks which have specific start and end dates and which meet specific objectives.

prototyping: simulation of the real product.

record processor: a programming language which processes data one record at a time.

recovery: the process by which data is recovered after an unplanned interruption of program execution.

relationship: the connection of one object to another.

requirements: the processing and resourses needed to accomplish the objectives.

retention: the length of time that data is stored.

run time: the length of time required to executive the program code.

scope: boundary of the system or project.

security: restriction of access to the data for prevention of accidental or malicious modification, destruction, or disclosure.

set processor: a computer language that processes data a table at a time.

software: computer programs.

source: origin of system inputs.

specifications: detailed description of system inputs, outputs, and data, including formats.

statement: an action to be performed unconditionally.

structured analysis: the activity of deriving a structured model of the requirements for a system.

structured coding: a technique that employs a top-down refinement strategy to produce code built from a small set of constructs (sequence, decision, and iteration).

structured design: the development of a blueprint for construction of a computer system, having the same components and interrelationships among the components as the original problem.

structured programming: structured coding.

subsystem: a part of a system that can be designed and implemented independently of the other parts.

subtype object: an object which is functioning in a different role than other objects of the type; for example, SAVINGS ACCOUNT CUSTOMER and LOAN CUSTOMER might be two different subtypes of the object BANK CUSTOMER.

supertype object: an object that has subtypes.

system: any planned response to a need or requirement; a set of application-related manual and mechanized activies.

system development methodology: the particular steps to be performed and products to be produced to develop a system.

system input: data that enters the outer boundary of a system.

system output: data which leaves the outer boundary of a system.

system test: a trial performance of all the activities, manual and mechanized, within the project scope under actual operating conditions.

table handler: a mechanized process for retrieving, storing, and changing data that is stored in the computer.

techniques: the method or details of procedure essential to expertness of execution of any activity.

testing: executing part or all of a system with the intent of finding defects.

tight English: narrative that uses nouns that are in the data dictionary, a restricted set of verbs, and an indented format.

tools: mechanized aids for system development techniques.

transaction: any data that initiates by its content some process or series of processes.

transaction-centered: a system in which the major functions are comprised of processes that are initiated in response to specific transactions.

transform centered: a system in which the major functions are comprised of processes that transform input data into output data.

tuning: the activity of making modifications to the code or the design for the purpose of making the system more efficient.

user: a person who uses the system.

user's guide: a document explaining how to use the system.

volume: number of occurrences

Glossary of Symbols

The following is a glossary of symbols for data flow diagrams, information models, structure charts, and Nassi-Shneiderman charts as used in this book. Various drawing conventions can be used, and which variation is used is not important as long as the documentation is consistent. As analyst workbenches become more common, the choice of workbench will influence which symbols an organization chooses.

Data flow diagrams

Data flow diagrams are network diagrams showing the processes which must be performed to transform the system input data into the system output data, as well as the relationships among those processes. Error handling processes usually are ignored in data flow diagramming unless the error results in a system output.

Processes are indicated by circles or rounded rectangles with the name of the process inside the bubble, as in S.1.

S.1. Process.

Data flows are shown as arrows, with the point of the arrow showing the direction of the main flow of the data. Minor interactions are ignored. The name of the data flow is written next to the arrow, as in S.2. Data flows may be groups of data elements or a single data element. Unless necessary for clarity, data flows in and out of data stores do not need to be named until the lowest level data flow diagrams.

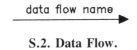

S.2. Data Flow.

Data stores are indicated by two parallel lines with the name of the data store between the lines, as shown in S.3.

DATA STORE NAME

S.3. Data Store.

External entities, sources of system inputs or recipients of system outputs, are shown as rectangles with the name of the external entity inside the box, such as in S.4. External entities can be people or other mechanized systems.

S.4. External Entity.

Data flow diagrams are used in several ways during the development process. The context diagram (Level 0 data flow diagram) is used early in the Requirements Phase.

The context diagram, S.5, includes all processes within the application being developed, both mechanized and manual. If the system is large, the inputs and outputs can be grouped according to (1) common data; and/or (2) a common external entity; and/or (3) common business event which triggers them.

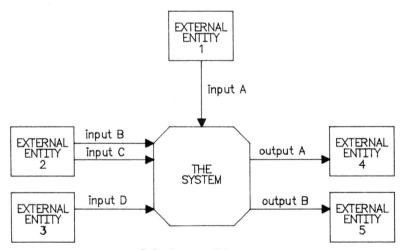

S.5. Context Diagram.

Event diagrams, S.6, are working documents used to help define the requirements. An event diagram can be thought of as a context diagram for one business event with its associated system inputs and outputs. Inputs to an event diagram must originate either from an external entity or from a data store, and its outputs must terminate at either an external entity or at a data store. Individual event diagrams are connected to give an overall view of the system. This diagram either will be exploded further and/or grouped, depending on its complexity and level of detail.

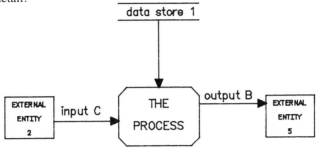

S.6. Event Diagram.

Leveled data flow diagrams are produced when a data flow diagram is partitioned (exploded) into more detail, such as in S.7.

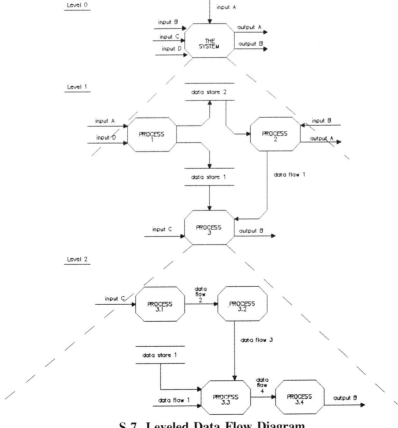

S.7. Leveled Data Flow Diagram.

When exploding data flow diagrams, each process input and output must show on the lower level diagram. No inputs or outputs may be added. Data flows will be added between subprocesses which were contained in the same higher level process. Also, a data flow may be subdivided into two new data flows so that some of the data elements can be input to one subprocess and the remainder into another subprocess. One major process does not need to be exploded into the same number of levels as another major process.

An expanded data flow diagram, S.8, is a consolidation of all the lowest level processes in a subsystem.

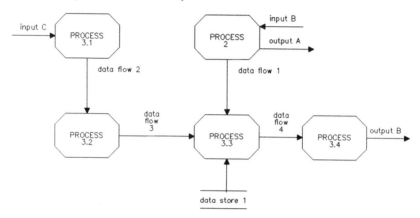

S.8. Expanded Data Flow Diagram.

Information models

An information model documents the objects that an organization stores data about and how those objects are related to each other. Objects in an object-relationship diagram are represented by a rectangle with the object name inside the box, as in S.9.

object name

S.9. Object.

Relationships are shown by a diamond with the relationship name inside the diamond, as in S.10.

S.10. Relationship.

Objects and relationships are connected with a straight line. An associative object is indicated by an arrow pointing toward the relationship. See S.11.

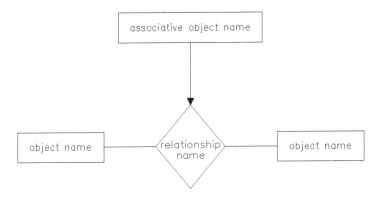

S.11. Object-Relationship.

Structure charts

A structure chart is a hierarchical arrangement of the processes in a subsystem. It includes all the processes which are in the expanded data flow diagram plus control, file access, table handling, formatting, and error handling processes. Each process is represented by a rectangle with the name of the process inside the box, as shown in S.12.

S.12. Process.

Subordinate processes are drawn beneath the processes that control their execution and are connected to them by a straight line, as in S.13.

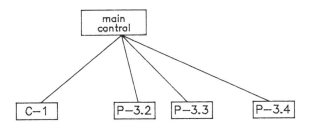

S.13. Subordinate Process.

LEGEND

C = Control Module
P = Process
I = System Input
O = System Output
DS = Data Store
DF = Data Flow

Structure charts may be drawn in leveled diagrams as in S.14

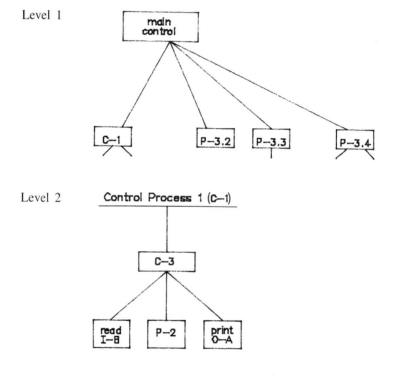

S.14. Leveling of Diagram.

Flow of data is shown by arrows with open-circle tails. Flow of control is shown by arrows with solid-circle tails. A circle with an arrow head indicates repetitive execution of the processes. A small diamond is used to show that one of the marked processes is chosen for execution. See S.15.

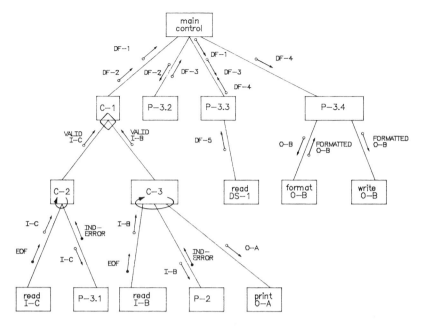

S.15. Structure Charts.

Nassi-Shneiderman charts

Nassi-Shneiderman charts are used for code design. Sequence is indicated by a rectangle, as in S.16, and iteration is drawn as in S.17.

Action 1
Action 2

S.16. Sequence.

Do until condition x
Action 1
Action 2

S.17. Iteration

Decision is indicated in two ways. One is a binary decision, S.18, and the other type a case, S.19.

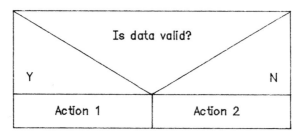

S.18. Binary Decision.

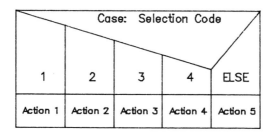

S.19. Case Decision.

One N/S chart should not extend to more than one page. If a process is referenced whose detail is on another page, the process name is circled, as in S.20.

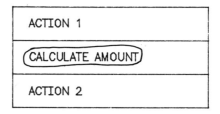

S.20. Off-page process.

Index

A

Accumulated Needs, 188
Add Curriculum Course, 152
Add Enrollment, 159
Add Group, 169, 182
Add Need, 154
Add Scheduled Course, 156
Add Student, 169, 184
Add Withdrawal Reason, 159
Ad-hoc reports, 225
Analysis, 225
Application, 225
Associates, relationship definitions, 140
Associative object, 225
Attribute, 225

B

Backup, 225
Binary decision, symbols used for, 238
Business data, 13, 225, 226
Business objectives, 128

C

Case, 225
  case decision, symbols used for, 238
Catalog, 134

Central transform, 225
Change Curriculum Course, 152
Change Group, 169, 183
Change Scheduled Course, 157
Change Student, 169, 184
Characteristic object, 225
Chargeback, 225
Code design, 182
  Add Group, 182
  Add Student, 184
  Change Group, 183
  Change Student, 184
  Delete Group, 183
  Delete Student, 185
  packaged model, 112
  route personnel transaction, 182
Cohesion, 225
Completion Date, 188
  information model, 19, 27
Completion Form, 121
Completion Report, 121, 134
Consolidated event diagram, 50-53
Constraint, 225
Context diagram, 130, 225, 232
Control flow, symbols used for, 236
Control function, 225
Conversion, 225
Coupling, 225
Course, object definitions, 139
Course Cost, 188
Course Days, 188
Course Description, 122, 188
  information model, 19, 27
Course End Date, 188
  information model, 19, 28
Course Level, 188
  information model, 19, 28
Course Name, 188
  information model, 19, 28
Course Number, 188
  information model, 19, 28

Course Prerequisite, 188
  information model, 19, 28
Course Roster, 122
Course Schedule, 122
Course Start Date, 188
  information model, 19, 29
Course Time, 189
  information model, 20, 29
Credit Hours, 189
  information model, 20, 29
Curriculum courses, 142
Curriculum Information, 189

D

Data, 226
Database, 226
Data conversion, 128
Data dictionary, 13-18
  business data, 13, 226
  data items, 13-17
  definition of, 13
  definitions, 188-91
    Accumulated Needs, 188
    Completion Date, 188
    Course Cost, 188
    Course Days, 188
    Course Descriptions, 188
    Course End Date, 188
    Course Level, 188
    Course Name, 188
    Course Number, 188
    Course Prerequisite, 188
    Course Start Date, 188
    Course Time, 189
    Credit Hours, 189
    Curriculum Information, 189
    Enrollment, 189
    Enrollment Date, 189

Group Information, 189
Group Manager Name, 189
Group Number, 189
Ind-End, 189
Instructor Name, 189
Message, 189
Needed Date, 189
Number of Needs, 189
Personnel Information, 189
Room Location, 189
Scheduled Course
 Information, 190
Selection Code, 190
Student Days, 190
Student Information, 190
Student Level, 190
Student Name, 190
Student Number, 190
Withdrawal, 190
Withdrawal Reason, 190
Sample Date Element
 Definition, 191
Sample Date Flow Definition,
 191
Data flow diagram, 226
 symbols used for, 231-34, 236
  data flows, 231
  data stores, 232
  external entities, 232
  processes, 231
Data items, 13-17, 226
 attributing data items to objects,
  27-33
Data stores, 226
 descriptions of, 142-44
  curriculum courses, 142
  group information, 142
  registration, 144
  scheduled courses, 143
  student information, 143
Debug, 226
Decision chart, 226

Decision tree, 226
Delete Curriculum Course, 152
Delete Group, 169, 183
Delete Need, 155
Delete Registration, 160
Delete Scheduled Course, 157
Delete Student, 169, 185
Derived data, 226
Design, 226
Design curriculum, 68-71, 151-52
 Add Curriculum Course, 152
 Change Curriculum Course, 152
 Delete Curriculum Course, 152
 Print Course Catalog, 152
 Update Curriculum Records,
  151
Design model, 92-104
 expanded data flow diagram, 95
 omission correction, 103-4
 structure chart, 95-103
  construction of, 95-101
  evaluation of, 102-3
 procedural analysis of, 219
Design model (maintain personnel
 information), 168-91
 code design, 182
  Add Group, 182
  Add Student, 184
  Change Group, 183
  Change Student, 184
  Delete Group, 183
  Delete Student, 185
  route personnel transaction,
   182
 module design, 179
 program design, 179
 report design layout, 177
  personnel list, 177
 screen design layout
  Add Group, 169
  Add Student, 169
  Change Group, 169

Design model (*cont.*)
  screen design layout (*cont.*)
    Change Student, 169
    Delete Group, 169
    Delete Student, 169
    personnel file menu, 168
  subsystem design structure
    chart, 178
Detailed process model, 67-91
  exploding the data flow
    diagram, 67-87
    Design Curriculum, 68-71
    Maintain Personnel
      Information, 83-87
    Schedule Training, 71-77
    Train Students, 77-83
  process allocation, 90
  revision of, 88-89
  walkthrough, 88
Development, 226
  development center, 226
Diagramming events, 46-50, 53
Documentation, 226

E

Efficiency, 226
Efficiency tuning, packaged
  model, 112
Enrollment, 189
Enrollment Date, 189
  information model, 19, 29
Enrollment Request, 123
Entity, 226
Event diagram, 232-33
Event list, 131
Events, 5-6
Executed, 226
Expanded data flow diagram, 226
Exploded data flow diagram, 226

F

Factored, 226
Feasibility, 226
File, 226
File access, 226
Forecasted need, 132
Forecasted Needs Report, 123
Format, 226
Fourth-generation language,
  65-66, 226
Frequency, 227
Function, 227
*Fundamental Concepts of
  Information* (Flavin), 2

G

Group, object definitions, 139
Group information, 142, 189
Group Manager Name, 189
  information model, 20, 30
Group Number, 189
  information model, 20, 30

I

Implementation, 227
Ind-End, 189
Information model, 19-33, 138-41
  Completion Date, 19, 27
  connection of objects and
    relationships, 235
  Course Description, 19, 27

Index

Course End Date, 19, 28
Course Level, 19, 28
Course Name, 19, 28
Course Number, 19, 28
Course Prerequisite, 19, 28
Course Start Date, 19, 29
Course Time, 20, 29
Credit Hours, 20, 29
Enrollment Date, 19, 29
Group Manager Name, 20, 30
Group Number, 20, 30
Instructor Name, 20, 30
Needed Date, 20, 31
No-Show Reason, 20, 31
object definitions, 139
    Course, 139
    Group, 139
    Needed Course, 139
    Registration, 139
    Scheduled Course, 139
    Student, 140
object-relationship diagram, 138
objects, 21-24, 234
    identification of, 24-27
relationship definitions, 140-41
    Associates, 140
    Needs, 141
    Works In, 141
relationships, 21-24, 234
Room Location, 20, 21
Student Level, 20, 31
Student Name, 20, 32
Student Number, 20, 32
Information modeling, xiv, 227
first-cut information model, 36-41
    adding associative object, 38
    attributing derived data, 39-41
    eliminating multiple relationships, 38
    eliminating objects, 37
refined information model, 41-44
    relationship and reverse relationship, 43-44
    Withdrawal Reason, 41
revision of, 116
Information requirements, 227
Instructor Name, 189
    information model, 20, 30
Interactive system, 227
Involved, 227
Iteration, symbols used for, 237

L

Level 1 Data Flow Diagram, 145, 150
Leveled data flow diagram, 233
List Accumulated Needs, 155
Logical, 227
Loop, 227

M

Maintain Personnel Information, 83-87, 162- 65
    Maintain Group Information, 163
        Add Group, 163
        Change Group, 163
        Delete Group, 164
    Maintain Student Information, 164-65
        Add Student, 165
        Change Student, 165
        Delete Student, 165
    Print Personnel List, 165

Maintain Personnel Information (*cont.*)
  process allocation, 166
Maintenance, 227
Maintenance request, 114-18
  analyzing change, 115-16
  revising information model, 116
  revising process model, 117-18
Message, 189
Mini-specs, 227
Model, 227
Module, 227
Module design, 179

N

Nassi-Shneiderman charts, 227
  off-page process, 238
  symbols used for, 237-38
    binary decision, 238
    case decision, 238
    iteration, 237
    sequence, 237
Needed Course, object definitions, 139
Needed Date, 189
  information model, 20, 31
Needs, relationship definitions, 141
Needs Form, 124, 135
No-Show Reason, information model, 20, 31
Number of Needs, 189

O

Object, 227
Object definitions, 139
  Course, 139
  Group, 139
  Needed Course, 139
  Registration, 139
  Scheduled Course, 139
  Student, 140
Objective identification, 5-6
Object-relationship diagram, information model, 138
Objects, information model, 21-24
Off-page process, Nassi-Shneiderman charts, 238
Omission correction, 103-4
On-line system, 227

P

Packaged model, 106-12
  code designing, 112
  efficiency tuning, 112
  grouping modules into programs, 107
  grouping processes into modules, 107
  procedural analysis, 108
Packaging, 227
PC, 227
Personnel file menu, 168
Personnel Information, 132, 189
Personnel list, 135, 177
Phase, 227
Physical implementation, 227
Print Course Catalog, 152
Print Needs Form, 153
Print Roster, 160
Print Schedules, 157
Problem definition, 128
Problem identification, 5-6
Procedural analysis, 228
  packaged model, 108
Process, 228

Process allocation, 228
  detailed process model, 90
Process description, 228
Processes, symbols used for, 235
Process model, 46-53, 145-47
  consolidated event diagram, 50-53
  diagramming events, 46-50, 53
  Level 1 Data Flow Diagram, 145
  process notes, 146-47
    Design Curriculum, 146
    Maintain Personnel Information, 147
    Schedule Training, 146
    Train Students, 146
  revision of, 117-18
Program, 228
Program design, 179
Project, 228
Projectscope
  events, 5-6
  objective identification, 5-6
  problem identification, 5-6
  sample forms and documents, 9-11
  system inputs, 6-7
  system outputs, 7-8
  system users, 7
Prototyping, 228

R

Record processor, 228
Recovery, 228
Refined process model, 55-66
  changing plans, 64
  defining information requirements, 62-63
  describing major porcesses, 59-62
  fourth-generation language, 65-66
  walking through event diagram, 55-58
Registrations, 144
  object definitions, 139
Relationship, 228
Relationship definitions, 140-41
  Associates, 140
  Needs, 141
  Works In, 141
Relationship and reverse relationship, 43-44
Report design layout, 177
Requirements, 228
Retention, 228
Room Location, 189
  information model, 20, 21
Roster, 136
Route personnel transaction, 182
Run time, 228

S

Sample forms and documents, 9-11
Scheduled Course, object definitions, 139
Scheduled Course Information, 190
Scheduled courses, 143
Schedules, 136
Schedule Training, 71-77, 153-57
  List Acculumated Needs, 155
  Print Needs Form, 153
  Print Schedules, 157
  Update Needs, 154-55
    Add Needs, 154
    Delete Need, 155
  Update Schedule, 156-57
    Add Scheduled Course, 156

Schedule Training (*cont.*)
  Update Schedule (*cont.*)
    Change Scheduled Course, 157
    Delete Scheduled Course, 157
Scope, 228
Screen design layout
  Add Group, 169
  Add Student, 169
  Change Group, 169
  Change Student, 169
  Delete Group, 169
  Delete Student, 169
  personnel file menu, 16
Security, 228
Sequence, symbols used for, 237
Set processor, 228
Skill Requirement, 133
Software, 228
Source, 228
Specifications, 228
Statement, 228
Statistics, 137
Structure charts
  construction of, 95-101
  designing model, evaluation of, 102-3
  drawn in leveled diagrams, 236
  procedural analysis of, 219
  symbols used for, 235-37
    control flow, 236
    data flow, 236
    drawn in leveled diagrams, 236
    processes, 235
    subordinated processes, 235-36
Structured analysis, xiii-xiv, 228
Structured Analysis and System Specification (DeMarco), 3
Structured coding, xi-xii, 228

Structured design, xii-xiii, 229
Structured programming, 229
Structured techniques
  history of, xi-xiii
    information modeling, xiv
    problems with learning, xiv-xv
    structured analysis, xiii-xiv
    structured coding, xi-xii
    structured design, xii-xiii
Student, object definitions, 140
Student information, 143
Student Level, information model, 20, 31
Student Name, information model, 20, 32
Student Number, information model, 20, 32
Subordinated processes, symbols used for, 235-36
Subsystem, 229
Subsystem design structure chart, 178
Subtype object, 229
Supertype object, 229
System, 229
System development methodology, 229
System inputs, 6-7, 229
  requirements for, 132-33
    enrollment, 132
    forecasted need, 132
    Personnel Information, 132
    Skill Requirement, 133
    Withdrawal, 133
System objectives, 128
System out
  requirements for, 134-37
    catalog, 134
    completion reports, 134
    Needs Form, 135
    personnel list, 135

Index   247

roster, 136
schedules, 136
statistics, 137
System outputs, 7-8, 229
System overview, 128-29
   business objectives, 128
   data conversion, 128
   problem definition, 128
   system objectives, 128
   user needs, 129
System test, 229
System users, 7

T

Table handler, 229
Techniques, 229
Testing, 229
Tight English, 229
Tools, 229
Training statistics, 125
Train Students, 77-83, 158-60
   Add Completion Date, 160
   Print Completion Report, 160
   Print Roster, 160
   Print Statistics, 161
   Update Registration
      Information, 159
      Add Enrollment, 159
      Add Withdrawal Reason, 159
      Delete Registration, 160
Transaction, 229
Transaction-centered, 229
Transform centered, 229
Tuning, 229

U

Update Curriculum Records, 151
Update Needs, 154-55
   Add Need, 154
   Delete Need, 155
Update Registration Information,
   159
   Add Enrollment, 159
   Add Withdrawal Reason, 159
   Delete Registration, 160
Update Schedule, 156-57
   Add Scheduled Course, 156
   Change Scheduled Course, 157
   Delete Scheduled Course, 157
User, 229
User needs, 129
User's guide, 229

V

Volume, 229

W

Walkthrough, detailed process
   model, 88
Withdrawal, 133
Withdrawal reason, 41
Works In, relationship definitions,
   141